The Physics of
Actinide Compounds

PHYSICS OF SOLIDS AND LIQUIDS

Editorial Board: Josef T. Devreese • *University of Antwerp, Belgium*
Roger P. Evrard • *University of Liège, Belgium*
Stig Lundqvist • *Chalmers University of Technology, Sweden*
Gerald D. Mahan • *Indiana University, Bloomington, Indiana*
Norman H. March • *University of Oxford, England*

SUPERIONIC CONDUCTORS
Edited by Gerald D. Mahan and Walter L. Roth

HIGHLY CONDUCTING ONE-DIMENSIONAL SOLIDS
Edited by Jozef T. Devreese, Roger P. Evrard, and Victor E. van Doren

ELECTRON SPECTROSCOPY OF CRYSTALS
V. V. Nemoshkalenko and V. G. Aleshin

MANY-PARTICLE PHYSICS
Gerald D. Mahan

THE PHYSICS OF ACTINIDE COMPOUNDS
Paul Erdös and John M. Robinson

THEORY OF THE INHOMOGENEOUS ELECTRON GAS
Edited by S. Lundqvist and N. H. March

A Continuation Order Plan is available for this series. A continuation order will bring delivery of each new volume immediately upon publication. Volumes are billed only upon actual shipment. For further information please contact the publisher.

The Physics of
Actinide Compounds

Paul Erdös
University of Lausanne
Lausanne, Switzerland

and

John M. Robinson
Indiana University-Purdue University at Fort Wayne
Fort Wayne, Indiana

PLENUM PRESS • NEW YORK AND LONDON

Library of Congress Cataloging in Publication Data

Erdös, Paul, date —
  The physics of actinide compounds.

  (Physics of solids and liquids)
  Bibliography: p.
  Includes index.
  1. Actinide elements. I. Robinson, John M., date — . II. Title. III. Series.
  QD172.A3E73   1983   546'.4   83-2331
  ISBN 0-306-41150-4

© 1983 Plenum Press, New York
A Division of Plenum Publishing Corporation
233 Spring Street, New York, N.Y. 10013

All rights reserved

No part of this book may be reproduced, stored in a retrieval system, or transmitted,
in any form or by any means, electronic, mechanical, photocopying, microfilming,
recording, or otherwise, without written permission from the Publisher

Printed in the United States of America

Preface

The authors' aim is to present a review of experimental and theoretical research that has been done to establish and to explain the physical properties of actinide compounds. The book is aimed at physicists and chemists. It was thought useful to collect a large selection of diagrams of experimental data scattered in the literature. Experiment and theory are presented separately, with cross-references.

Not all work has been included: rather, typical examples are discussed. We apologize to all researchers whose work has not been quoted. Since we report on an active field of research, clearly the data and their interpretation are subject to change.

We benefitted greatly from discussions with many of our colleagues, particularly with Drs. G. H. Lander and W. Suski. The help of Mrs. C. Bovey and Ch. Lewis in the preparation of the manuscript, and the artwork and photographic work of Ms. Y. Magnenat and E. Spielmann of the Institute of Experimental Physics of the University of Lausanne, are gratefully acknowledged. Our particular thanks are due to Ms. J. Libby for her skillful and patient editorial work.

We express our thanks to the Swiss National Science Foundation and the Herbette Foundation of the University of Lausanne, who promoted the cooperation of the authors.

Contents

List of Tables   xi

CHAPTER 1. Introduction   1

CHAPTER 2. Survey of Experimental Data   7

2.1. Experimental Techniques 7
     2.1.1. Sample Preparation 7
     2.1.2. Neutron Diffraction........................... 8
     2.1.3. Nuclear Magnetic Resonance 9
     2.1.4. Mössbauer Resonance 9
     2.1.5. Muon Spin Rotation (μSR) 11
     2.1.6. Other Experimental Techniques 12
2.2. NaCl-Type Metallic Actinide Compounds 13
     2.2.1. General 13
     2.2.2. Uranium Monophosphide 16
     2.2.3. Uranium Monoarsenide 22
     2.2.4. Uranium Mononitride 26
     2.2.5. Uranium Antimonide......................... 31
     2.2.6. Uranium Monochalcogenides 33
     2.2.7. Neptunium Carbide 36
     2.2.8. Neptunium Monopnictides 38
     2.2.9. Plutonium Compounds....................... 42
     2.2.10. Solid Solutions of Uranium Monopnictides and Monochalcogenides 44
2.3. UX_2-Type and UXY-Type Tetragonal Uranium Compounds 51
2.4. A_3X_4-Type Metallic Actinide Compounds 58
2.5. Intermetallic Actinide Compounds 62
     2.5.1. General 62
     2.5.2. AX_2-Type Intermetallics 63
     2.5.3. AX_3-Type Intermetallics 69
     2.5.4. Miscellaneous Intermetallics 72

2.6. Actinide Oxides	74
2.6.1. Uranium Dioxide	76
2.6.2. UO_2–ThO_2 Solid Solutions	79
2.6.3. Other Uranium Oxides	81
2.6.4. Neptunium Dioxide	83
2.6.5. Intermetallic Oxides	85
2.7. Actinide Halides	86
2.7.1. Uranium Triiodide	86
2.7.2. Other Actinide Halides	90
2.8. Other Compounds	91
2.9. Spectroscopic Data on Actinide Ions	92

CHAPTER 3. Survey of Theory   **99**

3.1. Localized Electron Theories	99
3.1.1. Assumptions of the Crystal-Field Model	99
3.1.2. Crystal-Field Theory	103
3.1.2a. Mathematical Formulation	105
3.1.2b. The Coulomb Hamiltonian	106
3.1.2c. The Spin–Orbit Hamiltonian	108
3.1.2d. The Crystal-Field Hamiltonian	111
3.1.2e. Point Symmetry	113
3.1.2f. Matrix Elements of $\mathcal{H}_{cf}$	115
3.1.2g. The Zeeman Hamiltonian	117
3.1.2h. Thermodynamic Averages	117
3.1.2i. Shielding	120
3.1.2j. Comments on the Crystal-Field Theory	121
3.1.3. Interionic Interactions	123
3.1.3a. Heisenberg Exchange	123
3.1.3b. Anisotropic Exchange	124
3.1.3c. Electric Multipole Interactions	125
3.1.3d. Biquadratic Exchange	125
3.1.3e. RKKY Exchange	125
3.1.3f. Coqblin-Schrieffer Exchange	129
3.1.4. Theories of First-Order Transitions Based on Localized Models	133
3.1.4a. Biquadratic Exchange	133
3.1.4b. Allen's Theory of UO_2	135
3.1.4c. Theories of NpO_2	142
3.1.4d. The Long–Wang Theory of UP	144
3.1.4e. Blume's Level-Crossing Theory	146
3.1.4f. A Theory of UI_3	147
3.2. Theories Involving Itinerant Electrons	148

3.2.1. Band Calculations 149
　　　　3.2.1a. Introduction 149
　　　　3.2.1b. Actinide Metals 150
　　　　3.2.1c. NaCl-Type Actinide Compounds 152
　　3.2.2. Spin-Fluctuation Models 155
　　3.2.3. Electron Delocalization Model 157
　3.3. A Theory Intermediate to the Localized and Band Models 164
　3.4. Similarities between Mixed-Valence Rare Earths and Metallic Actinide Compounds 169

APPENDIX A. Irreducible Tensor Operators 171

APPENDIX B. Magnetic Structures 175

References 185

Author Index 199

Subject Index 207

List of Tables

1. The Three Groups of Magnetic Elements, and a List of the Actinide Elements Together with Their Electronic Configurations outside the [Rn] Core — 2
2. The Major Classes of Actinide Compounds — 3
3. Compounds of the Type AX, with the NaCl Structure — 14
4. Tetragonal Conductor or Semiconductor Compounds of the Type AX_2 and AXY — 52
5. A_3X_4-Type Compounds of Body-Centered Cubic Structure — 59
6. Intermetallic Actinide Compounds of the Type AX_2 — 64
7. Intermetallic Actinide Compounds of the Type AX_3 — 65
8. Selected Properties of Actinide Oxides — 75
9. Listing of Actinide Halides — 86
10. D_{2d} Spectroscopic Parameters of U^{4+} — 93
11. Crystal-Field Parameters and Mean Radii of Actinides — 95
12. Observed and Calculated Crystal-Field Splittings of the Ground State of Np^{3+} in $LaCl_3$ — 96
13. Slater Integrals F^i and Spin–Orbit Coupling Parameter λ for Trivalent Actinides in $LaCl_3$ Host — 97
14. Comparison of Experimental and Theoretical Values of the Slater Integrals F^i and Spin–Orbit Coupling Constant for U^{3+} Ions in $LaCl_3$ — 97
15. Composition of the Eigenstates of the Lowest-Lying Manifold of Crystal-Field States of the U^{3+} Ion, as Observed in $LaCl_3$ Host — 109
16. Sternheimer Shielding Factors for Uranium Ions — 112
17. General Form of the Crystal-Field Potential for Electrons with $l \leqslant 3$ — 114
18. Anisotropic Interaction Constants for Ce^{3+} Pairs in $LaCl_3$ and $LaBr_3$ — 124
19. Theoretical Parameters and Predictions of Allen's Theory of UO_2 — 141
20. Parameters of Several Theories of Actinide Spin–Fluctuation Systems — 156

CHAPTER 1

Introduction

Most research into the magnetic properties of crystalline solids in the last forty years has been focused on the iron-group transition metals, the rare earth metals (such as gadolinium), and the vast number of compounds of these metals with other elements. It is for the most part only in the last fifteen years that the third group of magnetic elements, the actinides, have come under experimental and theoretical investigation (see Table 1). The actinides are elements with atomic number greater than 89, of which U, Np, and Pu are the best-known examples. They have many radioactive isotopes requiring special care in handling. The compounds of actinide elements include metals, semiconductors, and insulators and fall into the eight broad classes listed in Table 2. These materials are usually either ferromagnetically or antiferromagnetically ordered at low temperatures.

Interest in actinide compounds has been stimulated by the observation of some quite unusual magnetic, electronic, and thermodynamic properties, the most striking of which are first-order magnetic phase transitions found in materials belonging to classes one, six, and eight of Table 2. These transitions involve discontinuous changes in the ordered magnetic moment per actinide ion as a function of temperature, and some examples are shown schematically in Fig. 1. The behavior in parts (B) and (C) of Fig. 1 obviously contrasts strongly with the "normal" Brillouin curve of part (A) observed most often in the iron and rare earth group compounds where the ordered moment vanishes continuously as T is raised through the Néel temperature T_N or the Curie temperature T_C. These transitions are associated with the following properties:

1). At a certain temperature T' ($< T_N$ or T_C), the ordered moment σ per actinide ion decreases continuously by $\sim 10\%$ with increasing T. Examples of compounds exhibiting this "moment-jump" transition are the metals UP,[1] UAs,[2,3] and NpC.[4]

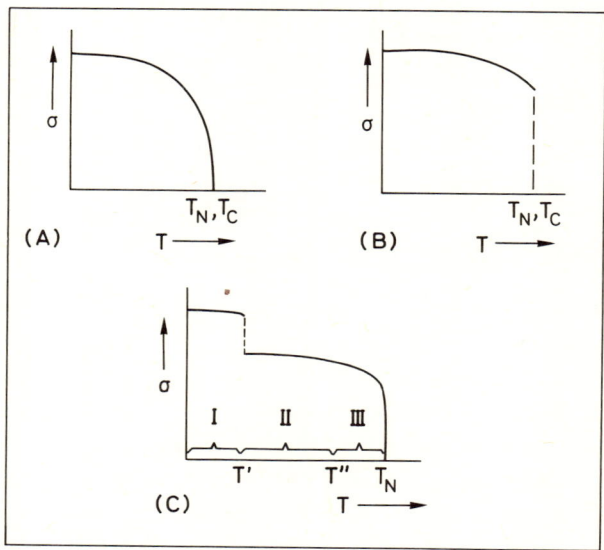

Fig. 1. Schematic illustration of types of magnetic phase transitions. Solid curve: average magnetic moment σ per ion vs. temperature T. (A) Second-order phase transition from ordered to paramagnetic phase at Néel (T_N) or Curie (T_C) temperature, observed most often in the iron and rare-earth groups. (B) and (C) First-order phase transition at temperatures T', T'', T_N, and T_C observed in the actinide group. I, II, and III: Temperature regions of possibly different magnetic order (symmetry).

Table 1. The Three Groups of Magnetic Elements, and a List of the Actinide Elements Together with Their Electronic Configurations outside the [Rn] Core

	Group		Example		Configuration		
	Transition metals		Fe		[Ar] $3d^6 4s^2$		
	Rare earths		Gd		[Xe] $4f^7 5d^1 6s^2$		
	Actinides		U		[Rn] $5f^3 6d^1 7s^2$		
Ac	Th	Pa	U	Np	Pu	Am	Cm
$6d^1 7s^2$	$6d^2 7s^2$	$5f^2 6d^1 7s^2$	$5f^3 6d^1 7s^2$	$5f^5 7s^2$	$5f^6 7s^2$	$5f^7 7s^2$	$5f^7 6d^1 7s^2$
Bk		Cf	Es	Fm	Md	No	Lr
$5f^7 6d^2 7s^2$		$5f^9 6d^1 7s^2$	—	—	—	—	—

Table 2. The Major Classes of Actinide Compounds

Class	Examples	Properties
NaCl-type	UP, PuC, NpN	Metallics (largest class)
UX_2 and UXY	UP_2, UOS	Tetragonal conductors and semiconductors
U_3X_4	U_3P_4	Body-centered cubic metallic ferromagnets
U_2N_2M	M = P, As, S, Se	Hexagonal antiferromagnets
U_2N_2Z	Z = Sb, Bi, Te	Tetragonal ferromagnets
Oxides	UO_2, NpO_2	Cubic CaF_2-type insulators
Intermetallics	$NpAl_2$, $NpPd_3$	Cubic
Halides	UI_3	Orthorhombic or hexagonal insulators

2). The moment-jump transition may (e.g., in UAs, NpC) or may not (e.g., in UP) be accompanied by a change in the magnetic symmetry, i.e., in the type of magnetic ordering, of which several examples are shown in Fig. 4 (p. 15).

3). The continuous portions of the sublattice magnetization curve do not in general follow a Brillouin function. For instance, at T_N the magnetization vanishes quite steeply for UP and UAs.

4). The magnetic susceptibility usually increases sharply by a factor of about two at the transitions at $T = T'$ and $T = T_N$.[5,6]

5). The electrical resistivity increases suddenly at $T = T'$ [6,7] as well as at $T = T_N$.

6). There are extremely sharp peaks in the specific heat vs. T at these transitions.[8]

7). There are at most small ($\sim 10^{-4}$) changes in the volume and no change in the lattice structure at the first-order transitions.[4,9]

8). At a temperature T'' ($< T_N$ or T_C) the magnetic symmetry may change suddenly without any significant change in the magnetic ordered moment per actinide ion (example: $UP_{0.75}S_{0.25}$).[10]

9). In the insulators UO_2[11] and UI_3[12] there are no moment-jump transitions at T', but the magnetization drops discontinuously from 80–90% of its maximum value to 0 at $T = T_N$.

10). In UAs no critical neutron scattering occurs at the phase transition temperature T'.[2] (As a rule, critical scattering indicates a second-order transition and its absence points to a first-order transition.)

Considerable theoretical research has been devoted to trying to explain these anomalous transitions, and a number of physical models have been proposed.[13-16] A study of the proposed models is particularly rewarding, because in dealing with the complexity of the actinides one is forced to question the underlying assumptions of many of the customary theories of magnetism which we often take for granted, such as the *crystal field* and *Heisenberg exchange* models. These complications arise from the special characteristics of the $5f$ electrons that are primarily responsible for the magnetism of the actinides, as compared with the $3d$ and $4f$ electrons of the iron-group transition metals and the rare earths, respectively.

A precise knowledge of the electronic structure of the actinide compounds is at present lacking. There is sufficient evidence, however, to show that the electronic band structures of the actinide elements and their metallic compounds involve overlapping $5f$, $6d$, and $7s$ bands, so that some of the $5f$ electrons are neither well localized on the actinide ions nor completely itinerant. Band structure calculations are complicated by the need to take into account electronic correlation in the narrow f bands. Thus, the actinide compounds may provide a testing ground for the theories of the role of electronic correlation in magnetism, such as the *Hubbard* and *Anderson* models. These theories aim to explain the variation from localized moment behavior to itinerant magnetism and from insulator to conductor. Since there are actinide compounds that show such variations as either the temperature or pressure is changed,[17,18] there are rich possibilities for comparison of theory and experiment.

Because of the large number of experimental and theoretical results that has accumulated recently, it seems appropriate to place the picture in perspective with a comprehensive review. The physical properties of the NaCl-type metallic actinides known as of 1967 were reviewed by Grunzweig–Genossar *et al.*,[19] but much new data are available. More recent but shorter reviews are those of Brodsky,[7] who deals with the pure metals and metallic compounds, Holtzberg *et al.*,[20] and Trzebiatowski and Troć,[21,256] who have summarized work on intermetallic and ferromagnetic uranium compounds. Results accumulated between 1976 and 1978, with the explicit exclusion of monocompounds and dioxides, are reviewed by Suski.[22]

The excellent book *The Actinides – Electronic Structure and Related Properties*[23] contains an extensive tabulation of experimental and theoretical work done before 1973. The theoretical part of the book is concerned mostly with traditional band and crystal-field calculations. The book omits from its review theories of the first-order magnetic phase transition, one of the most interesting aspects of actinide compounds. The book's review of experimental data does not emphasize materials that have particularly interesting magnetic properties; for example the "moment-jump" transition in UP receives scant mention.

In the present work, we concentrate on the phase-transition viewpoint and give less attention to phenomena such as resistivity and susceptibility of the elemental metals and intermetallic compounds, photoemission spectroscopy, nuclear γ-ray, and quadrupole resonance, which are covered in sufficient detail in the above mentioned book. Furthermore, in the present work the various physical properties of each material are discussed together in the same section, whereas in the Reference twenty-three different properties are analyzed in different articles.

The papers presented at the 2nd and 3rd International Conferences on the Electronic Structure of the Actinides are published in References 204 and 407, respectively. The work presented at the International Conference on the Physics of Actinides and Related $4f$ Materials in 1980 is published in Reference 408; the latter papers have also been reprinted in book form in Reference 409.

In the next chapters, the experimental results concerning the six main classes of actinide compounds are reviewed and discussed. In Chapter 3 the earlier theories based on the assumption of $5f$ electrons well localized in the crystal field are described and evaluated. Theoretical attempts to elucidate the complex band structure of the metallic actinides and its role in the first-order transitions are presented. Finally, suggestions for new research are discussed both in the review section and in the theoretical part.

CHAPTER 2

Survey of Experimental Data

2.1. Experimental Techniques

2.1.1. Sample Preparation

Actinide compounds and their single crystals are produced in the following laboratories: Transuranium Research Laboratory in Oak Ridge National Laboratory, Oak Ridge, TN 37830, U.S.A.; Commission of the European Communities, Joint Research Center, European Institute for Transuranium Elements, Postfach 2266, 7500 Karlsruhe 1, Germany; Institute for Low Temperature and Structure Research, Polish Academy of Sciences, P.O.B. 937, 50-950 Wrocław, Poland; Laboratory for Solid State Research, Swiss Federal Institute of Technology, 8093 Zürich, Switzerland; Battelle Memorial Institute, Columbus, Ohio, U.S.A. This list is by no means complete, and many excellent samples are produced in other institutions as well.

A major obstacle to the understanding of actinide compounds has been the lack of large single crystals. Recently, several materials, such as UN,[24] UO_2,[25,26] NpO_2,[22] UFe_2,[27] UP,[28] UAs,[28] USb,[28] U_3P_4,[29] and a few other pnictides[26] of U and Pa have been made in monocrystalline form.

Radioactive or toxic actinides such as Pu metal must be handled in glove boxes with suitable safety precautions. Many compounds such as UI_3 must be kept at all times in an atmosphere of dry, inert gas to prevent decomposition or contamination.

The production of samples of actinide compounds for experiments usually begins with the pure actinide metal and often requires special techniques.[30]

For example, in the production of uranium phosphides (UP, UP_2, U_3P_4), the uranium metal is first changed to UH_3 powder by exposure to hydrogen gas at 230°C.[8] After introducing a flowing stream of a phosphine + argon gas mixture and heating gradually to 800°C, the UH_3 changes to UP_2. The phosphine is cut off, and the UP_2 decomposes into U_3P_4, which is pressed into a pellet to raise its density. By heating the U_3P_4 to 1400°C in a vacuum, UP in the form of a gray powder can be produced.

Polycrystalline compounds are often made[31] by placing stoichiometric proportions of the constitutent elements in sealed ampoules and heating them above the reaction temperature. Sometimes fairly large single crystals are obtained in this process.

Single crystals of U_3P_4, UAs, etc. were made by the chemical vapor transport method.[32-33] In this method, the polycrystalline compound is placed in one end of an ampoule (usually quartz), which is evacuated to 10^{-6} mm Hg pressure. A transporting gas (I_2, Br_2, or $TeCl_4$) is introduced, and the sealed ampoule is placed in a tube furnace having a temperature difference which varies from 50[33] to 1000°C[34] between its two ends. A reaction ensues, in which the actinide compound decomposes to form a volatile compound with the transport gas, for instance, an iodide. At the other end of the ampoule, which is at a higher or lower temperature (depending whether the reaction is exo- or endothermic), the inverse reaction takes place and single crystals of the original compound are grown on a substrate. The latter is either a resistance-heated wire, or a radio-frequency induction-heated metal substrate.

Classical methods, such as pulling from the melt or high-temperature solution growth are also applied to try to grow single crystals.[410]

In general, with the most successful methods, single crystals of a few millimeters were obtained.[29] Crystals of UO_2 with dimensions of several centimeters are available,[25] but even these samples contain many antiferromagnetic domains in the ordered state, each with a different orientation of the magnetic moments.

2.1.2. Neutron Diffraction

Neutron diffraction is a most important experimental tool for determining the magnitudes, spatial directions, and temperature dependencies of the magnetic moments of actinide ions in the ordered state. In the experiment, a beam of thermal neutrons is incident on a crystal (or powder sample) of the material to be studied, and the angular positions of the constructive reflections are recorded. If the crystal is magnetically ordered, one set of reflections results from Bragg scattering from the periodic lattice of all atoms in the crystal, exactly as in X-ray diffraction. A second set of reflections results from the interaction of the neutron spin with the magnetic moments of the magnetic

2.1. EXPERIMENTAL TECHNIQUES

ions. A neutron with its spin aligned in a given direction scatters with different amplitudes from two atoms whose respective magnetic moments have different orientations. The temperature dependencies of the intensities and the Miller indices of the magnetic reflections yield the temperature variation of the magnitudes and orientations of the ordered magnetic moments of the sublattices of magnetic ions. For further details concerning neutron diffraction see Appendix B. One may also consult Reference 35.

2.1.3. Nuclear Magnetic Resonance

Nuclear magnetic resonance (NMR) experiments may be carried out on compounds containing at least one type of atom with a nonzero nuclear spin (e.g., the ^{31}P nucleus in UP). In the magnetically ordered state, there arise large (~100 kG) internal magnetic fields at the nuclear sites due to the difference in the numbers of "spin-up" and "spin-down" electrons. The energy differences between the nuclear spin states of different magnetic quantum numbers will be proportional to the magnitude of the internal field at the site of the nucleus. The latter field is in turn usually proportional to the average magnitude of the ordered magnetic moments of the surrounding magnetic ions. Thus, if the sample is placed in a cavity exposed to radio frequency radiation, there occurs resonant absorption at frequencies corresponding to allowed transitions between nuclear spin states (see Fig. 2). By measuring the temperature variation of the frequency ω required to produce a given resonance, one may infer (see Fig. 2) the temperature variation of the relative ordered moment $\sigma = \mu(T)/\mu(T=0\text{ K})$ from the relation $\sigma = \omega(T)/\omega(T=0\text{ K})$. In the usual case the magnitude of $\mu(T)$ itself is not determined, because the coefficient of proportionality between the internal field and the ordered magnetic moment is usually unknown. The reader is referred to Reference 36 for further details of NMR techniques.

2.1.4. Mössbauer Resonance

In Mössbauer resonance experiments, low-energy γ-rays are emitted by excited nuclei (e.g., ^{237}Np) of atoms in a source (e.g., NpAl$_2$). These γ-rays are incident upon an absorbing material (e.g., NpOs$_2$) containing nuclei of the same type in their ground states. There is a certain probability for a nucleus to absorb or emit a γ-ray without recoiling, in which case the crystal as a whole recoils to conserve momentum. Since the kinetic energy of the recoiling crystal is negligible, energy conservation shows that the energy of the γ-ray is simply equal to the energy difference between the excited and ground nuclear states.

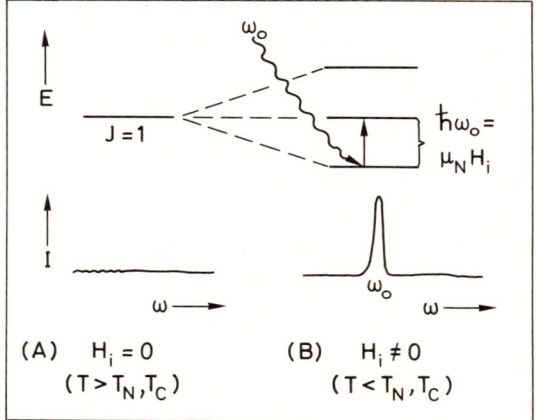

Fig. 2. Principle of nuclear magnetic resonance experiment to measure the relative magnitude (e.g., vs. T) of the magnetic field H_i at the site of a nuclear spin J in a solid. *Top*: The $J = 1$ level, degenerate in the absence of the field (A), is split by the field H_i (B). Electromagnetic radiation of frequency $\omega_0 = \mu_N H_i/\hbar$ (μ_N = nuclear magneton) is absorbed in a transition between two nuclear Zeeman levels. *Bottom*: Intensity I of the absorption vs. ω in cases (A) and (B). The field H_i is produced by ordered moments at temperatures $T < T_N$ (Néel) or $T < T_C$ (Curie).

The latter condition makes possible the resonant absorption of the γ-rays by the absorber. The resonance is usually observed as a loss in the number of γ-rays transmitted through the absorber as the energy of the γ-rays is varied by moving the source relative to the absorber with an adjustable velocity v ($v \sim 1$ mm/sec). Typical transmission spectra are shown in Fig. 3. If the absorber is magnetically ordered, the internal field H_i will split the energies of the spin sublevels of the nuclear states and give rise to additional resonances. As in the NMR technique, the magnitude of the ordered magnetic moments of the magnetic ions is assumed to be proportional to H_i and therefore to the separation between successive resonance energies. With this assumption the temperature dependence of the ordered moments may be inferred from the temperature dependence of the resonance energies.

If the physical environment of the absorbing nuclei differs from that of the emitted nucleus, a finite velocity δ is required to achieve resonance, even if the internal field H_i is zero (see Fig. 3). The quantity δ is the *isomer shift* and results from a difference between the density of the s electrons at the nucleus

2.1. EXPERIMENTAL TECHNIQUES

of the emitter and the s-electron density at the absorber nucleus. The s-electron density in turn depends on the number of electrons localized in the $5f$ shell. Thus, the decrease in the isomer shift is an indication of the localization of the $5f$ electrons of the absorber relative to those of the emitter material.

Mössbauer resonance experiments on actinide materials have been limited largely to compounds of neptunium, because uranium has no isotope with suitable energy levels. For further details of Mössbauer resonance and related techniques, see References 37, 38, and 255.

2.1.5. Muon Spin Rotation (μSR)

Intense beams of positive and negative muons (μ-mesons) are produced by particle accelerators. These mesons, which have the same charge as positrons

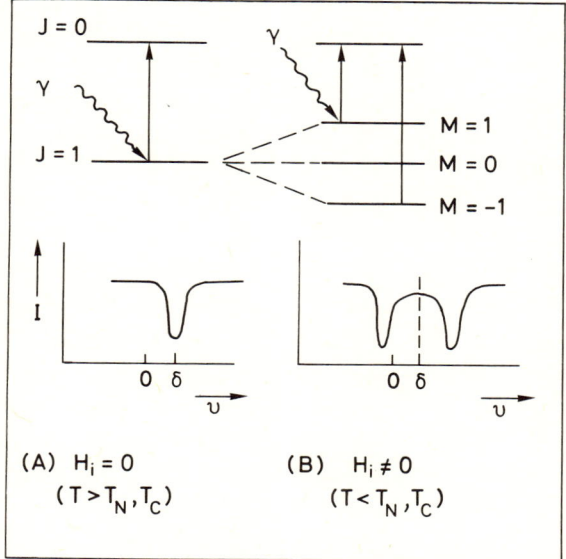

Fig. 3. Principle of Mössbauer resonance experiment to measure the relative magnitude of the magnetic field H_i (cf., Fig. 2). Top: Absorption of a γ-ray by a hypothetical nucleus with spin $J = 1$ in the absence (A) and in the presence (B) of the magnetic field H_i. Bottom: Intensity I of γ-rays transmitted through a sample of such nuclei as a function of the relative velocity v between source and absorber, which produces a Doppler shift of the absorption frequency. δ: isomer shift (see text).

and electrons, a spin of 1/2, and a mass $m_\mu = 207\, m_e$ are stopped inside the magnetic material. The muon beams are highly spin-polarized along their direction of motion, and their associated magnetic moment $g_\mu e/2m_\mu c$ begins to precess around the direction of the local magnetic field, until they decay into a positron (electron) and two neutrinos $[\tau(\mu^+) = 2.2\,\mu s]$. One observes the asymmetry in the angular distribution of the decay positrons (electrons) with respect to the direction of the incident muon beam. This asymmetry is a periodic function of time, the period being equal to the precession period of the muon magnetic moment. The latter is proportional to the local magnetic field and allows its determination.

Some particular features of μSR are 1). The asymmetry coefficient of μ^+ is five times larger than that for μ^-; 2). The μ^+ usually stops at interstitial positrons and may diffuse between such positions; and 3). The hyperfine field is determined by the spin density at the site of the muon and is generally much smaller than the hyperfine fields typical in NMR experiments, because the conduction electron spin polarization at interstitial sites is, as a rule, smaller than the s-electron spin polarization at a nuclear site. In addition, the hyperfine field is changed by the presence of the muon, since the latter causes a local change in the electron charge density (screening enhancement).

The smallness of the muon hyperfine field implies that the magnetic dipole field created by the ordered moments in the crystal represents an important contribution to the internal field.

The negative muons may form muonic atoms of small Bohr radii ($r_\mu = r_B/207$) with nuclei in the crystal, and during the lifetime of this atom, muon spin precession takes place. Again, the asymmetry of the muon decay spectrum yields information on the magnetic field within a sphere of radius r_μ. Hence μ^-SR may be used to probe the magnetic field at the site of spinless nuclei. The method of μ^-SR has been successfully used to probe iron- and rare-earth group compounds. Its application to actinide compounds is in progress.

2.1.6. Other Experimental Techniques

Among the conventional techniques, we mention X-ray diffraction, used for the determination of lattice parameters and structure. The magnetization and magnetic susceptibility are usually measured by the Gouy or Faraday methods, and using a mutual inductance bridge. Optical and electron-spin resonance spectroscopy are used to determine energy levels of actinide ions, including vibronic spectra.

High resolution X-ray photoemission spectroscopy is one of the more recent tools used to gain information on the electron energy distribution in actinide materials.[400, 404]

This method utilizes the photoelectric effect: An incident photon transfers its energy to an electron which is emitted. The knowledge of the energies of the incident photons and the measurement of the kinetic energy and the yield of the emitted electrons provide information about the density of states in the solid the electrons occupied before emission. It is customary to distinguish ultraviolet photoelectron spectroscopy (UPS) and X-ray photoemission spectroscopy (XPS) according to the photon source used.

Inverse photoelectron spectroscopy (IPE) is a method to obtain information about the density of states of the unoccupied electron states above the Fermi level. In this method electrons make radiative transitions to the unoccupied states and the energy of the resulting photons is measured. Depending on whether the energy of the incident electron is held constant and the photon energy is scanned, or *vice versa*, the method is called *bremsstrahlung isochromat spectroscopy* or *fluorescence IPE spectroscopy*, respectively. For more details and application, see References 429–432.

The electrical resistivity, thermopower, and specific heat of actinide ions are measured by conventional techniques.

2.2. NaCl-Type Metallic Actinide Compounds

2.2.1. General

The actinide compounds with the NaCl lattice structure have been extensively studied. They have the general formula AX, where A = actinide element and X = element of columns VA or VIA of the periodic table. Table 3 lists those compounds for which significant experimental data are known at present. These materials are fairly good conductors of heat and electricity, having room temperature specific resistivities ρ of a few hundred $\mu\Omega$ cm. Thus, one expects a partially filled conduction band. Exceptions are NpP and the semiconductor PuS, which have considerably higher resistivities. There are often anomalous variations of ρ and the thermoelectric power with temperature (see e.g., Figs. 11, 30, 34, 38[6,7,18,40-44]), especially near magnetic transition temperatures, indicating a coupling between magnetic and electronic behavior. One finds generally a high electronic contribution to the heat capacity,[8,19,45] which indicates high values of the density of electronic states at the Fermi level for the metallic actinide compounds. There are anomalies in the heat capacities

Table 3. Compounds of the Type AX (A = actinide), with the NaCl Structure. (a_0: lattice constant; magnetic order: FM = ferromagnet; AFM, AFM-I, AFM-IA: antiferromagnets (see Fig. 4); LW = longitudinal wave; T_t: magnetic phase transition temperature(s); μ_0: saturation moment per actinide ion; μ_p: paramagnetic moment per A ion; θ: paramagnetic Curie (Néel) temperature; ρ: resistivity; γ: electronic specific-heat constant)

Compound AX	a_0 (Å)	Magnetic order	T_t (K)	μ_0 (μ_B)	μ_p (μ_B)	θ (K)	ρ ($\mu\Omega$cm)	γ 10^{-4} $\frac{\text{cal}}{\text{m-deg}^2}$	Refs.
UAs	5.779	AFM-IA; I	63; 127	2.24	3.54	32	238	127	2, 19, 28
UBi	6.364	AFM-I?	285?	3.0?	4.06	115	—		19, 246, 247
UC	4.951	None	—	—	—	—	50	47	8, 46, 44
UN	4.890	AFM-I	53	0.75	2.7; 3.11	$-247; -325$	150; 160	96–62	19, 8, 46, 63, 65, 24
UP	5.589	AFM-I	22.5; 121	1.95	3.15	0	370	23	5, 8, 46, 1
US	5.489	FM	180	1.7	2.22	180	320	56	19, 72, 73
USb	6.197	AFM-I	213	2.82	3.64; 3.85	95; 140	357		19, 247, 69, 7, 70, 28, 267
USe	5.744	FM	160	2.0	2.51	180	600	208	19, 73
UTe	6.161	FM	108	2.2	2.84	108	1300		19, 73
NpAs	5.835	AFM	140; 175	~2.6	2.6	190			90, 78
NpC	4.992	FM; AFM-I	220; 310	2.2	3.22	225	200	101	4, 7, 5, 89
NpN	4.898	FM	87	1.4	2.44	100	400		5, 23, 90
NpP	5.610	LW	74; 120	2.3	2.8	125	3740		23, 90
NpS	5.527	FM	23	0.9	2.1	-150	1500		23, 248
NpSb	6.249	AFM?	205	~2.6	~2.3	150			23, 44
PuAs	5.86	FM	129	0.35	0.97	129			94
PuC	4.97	AFM?	60(?)	0.8?	2.37	55.4	257	31–48	23, 44, 293
PuN	4.91	AFM	13?	<0.2	1.1	-200		157	23, 294, 295
PuP	5.55	FM	126	0.42, 0.77	1.06	130	752	<27	93, 296, 297
PuS	5.537	None	—	—	—	—	2600		23, 41
PuSb	6.24	FM	85	0.57	1.0	90	—		23

2.2. NaCl-TYPE METALLIC ACTINIDE COMPOUNDS

at magnetic transition temperatures (Figs. 7, 16, 30), but the entropy decreases associated with magnetic ordering are often considerably smaller than those calculated by statistical mechanics under the assumption that an ensemble of localized magnetic moments orders.[48] These materials tend to be highly refractory, with melting points on the order 2500–3000 K.[41,47,48] Strong covalent bonding is indicated by these structural properties and by vaporization experiments which show evaporation of molecular units from the crystal,[49] as well as by the properties of the optical phonons as measured in high-resolution inelastic neutron scattering experiments.[50]

As Table 3 shows, most of these compounds order ferromagnetically or antiferromagnetically and have been investigated by neutron diffraction experiments. The magnitudes of the ordered moment μ_0 extrapolated to $T = 0$ K as determined by neutron diffraction are almost always markedly smaller than the corresponding paramagnetic moments μ_p deduced from applying the Curie–Weiss law[13] to the experimental paramagnetic susceptibility, as Table 3 shows. One should also notice that the inverse susceptibility $\chi^{-1}(T)$ often departs from the linear function of T predicted by the Curie–Weiss law (see e.g., Figs. 8, 21, and 22).

The three most common types of magnetic ordering found by neutron diffraction are the ferromagnetic (FM), antiferromagnetic type I (AFM-I), and antiferromagnetic type IA. These configurations are illustrated in Figs. 4 and 5. More complicated types of magnetic ordering are found in NpP[40] and in solid solutions such as $UP_{1-x}S_x$[10,51,52] and $UAs_{1-x}S_x$,[2] as will be described in detail later.

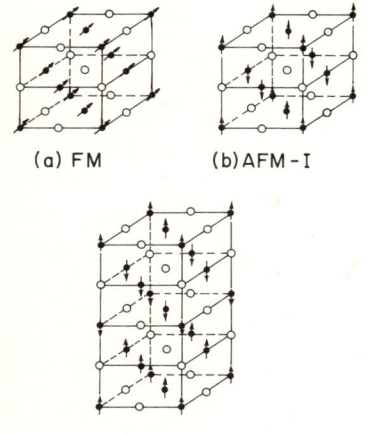

Fig. 4. Types of magnetic ordering in NaCl-type actinide compounds. Full circle: magnetic actinide ion. Empty circle: anion. Arrow: average direction of magnetic moment.

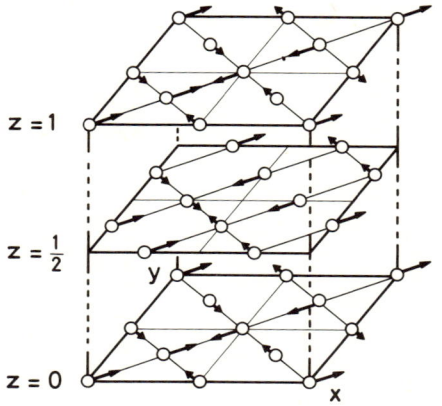

Fig. 5. Magnetic structure of UAs below 62 K, as proposed on the basis of a neutron diffraction experiment in Ref. 266. The open circles represent U-ions, and the arrows their magnetic moments which lie along ⟨1 1 0⟩ and equivalent directions.

We now turn to a consideration of a number of examples of special interest which quantitatively illustrate the above points.

2.2.2. Uranium Monophosphide

Uranium monophosphide, **UP**, is a Type I antiferromagnet with a Néel temperature $T_N = 121$ K,[5,8] at which temperature the magnetization becomes zero. The sublattice magnetization σ as a function of temperature T has been determined both by neutron diffraction[53] and by the much more accurate nuclear magnetic resonance (NMR) technique using powder samples.[1] Both NMR[1] and neutron diffraction[53,54,55] indicate a sudden drop in the ordered moment of about $0.2\,\mu_B$ upon heating to 22.5 K, as Fig. 6a shows. Because the intensities of all the observed neutron reflections and NMR lines simply decrease at $T' = 22.5$ K without the appearance of new reflections or lines,[53] one concludes[53] that the type of ordering remains AFM-I. The possibility of a rotation of the moments from the [001] to the [110] direction at T' was suggested[259] and recently verified by neutron measurements by Burlet *et al.*[411] on single crystals of UP in an applied magnetic field. Upon cooling of the sample from T_N to T' in a field of 25 kOe in the [110] directions, only the magnetic neutron reflections of the $\mathbf{k}_3 = [001]$ propagation direction of the collinear AFM-I structure are seen. At T', there is a sudden decrease in the $\mathbf{k}_3 = [001]$ reflection intensities and a sudden increase in the reflections corresponding to the $\mathbf{k}_1 = [100]$ and $\mathbf{k}_2 = [010]$ directions (see Fig. 6c). These data are interpreted by Burlet *et al.* to indicate a rotation of the U moments from the [001] direction for $T > T'$ to the [110] direction for $T < T'$. The magnetic ordering remains

2.2. NaCl-TYPE METALLIC ACTINIDE COMPOUNDS

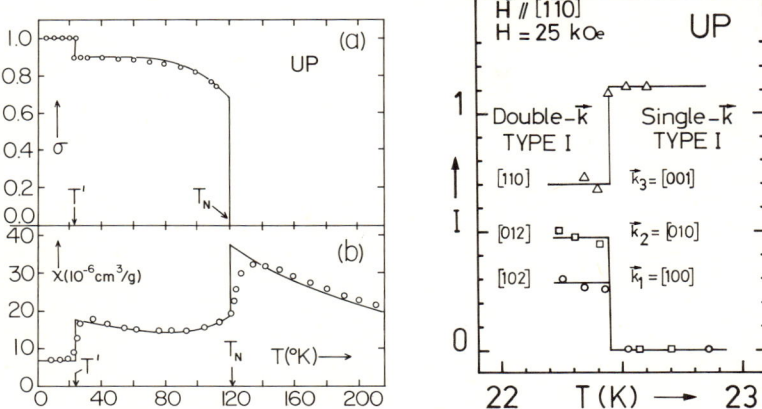

Fig. 6. a). Relative sublattice magnetization σ vs. temperature T in uranium phosphide, UP. The open circles represent experimental results obtained by nuclear magnetic resonance, Ref. 1. They coincide within experimental error with the neutron diffraction results of Ref. 153. b). Magnetic susceptibility of UP as a function of T. Experiment on powder samples of Ref. 5. Solid lines in (a) and (b): Theory of Ref. 16. On the right: Neutron diffraction intensity (in arbitrary units) vs. temperature in a UP single crystal in the neighborhood of the moment-jump transition temperature T'. The crystal is in a magnetic field of 25 kOe applied along the [110] crystallographic direction. One observes the appearance of different magnetic reflections below and above T', attributed to structures with k vectors as indicated. From Ref. 411.

Type I, the moment of each U atom being the sum of two components m_x and m_y along the k_1 and k_2 directions. This is an example of a "two-k" magnetic structure (see Appendix B). It is not yet clear, however, that the jump in the total U moment can be explained solely on the basis of the moment rotation.

Data show a sharp maximum at $T = T_N$ in the electrical resistivity $\rho(T)$.[56] The $\rho(T)$ behavior in UP near T_N is similar to that found in cubic $NpPd_3$, where the transition at $T = T_N$ is first-order.[57]

The sharpness of the transition at T_N is also indicated by the specific heat c_p vs. T data of Counsell et al.[8] (Fig. 7), which show that most of the entropy of magnetic disordering is produced within 10-15 K of the Néel temperature. The almost vertical ascent of the specific heat upon cooling to 121 K yields an accurate determination of the Néel temperature. The dramatic singularity in c_p at $T' = 22.5$ K corresponds to the sharp transition found at T' by the NMR experiment.[1]

The magnetic powder susceptibility $\chi(T)$ also increases steeply at $T = T'$ and $T = T_N$,[5] as shown in Fig. 6b. Other anomalous features are the shallow

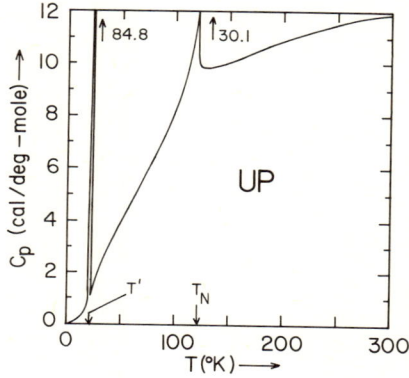

Fig. 7. Specific heat at constant pressure, c_p, of UP, experiments of Ref. 8. The numbers at the top are the peak values.

minimum in $\chi(T)$ at $T \simeq 90$ K, a small temperature-independent "plateau" between $T = 22$ K and $T = 23$ K, and the broad maximum above T_N.

The magnetic susceptibility has also been measured[28, 258] on single crystals of different orientations as a function of temperature in the range $T = 2$–1100 K (Fig. 8). These measurements in an applied field of 60 kOe confirm the picture obtained on the powder samples (Figs. 6b, 9). In addition, it is found that below T_N the observed moment for an applied field below ~ 120 kOe is obtained in the $\langle 100 \rangle$ direction, whereas in higher fields there occur sudden jumps in the magnetization, with $\langle 110 \rangle$ the easy axis (Fig. 10).[257] On powder samples and at lower fields hysteresis is observed.[58]

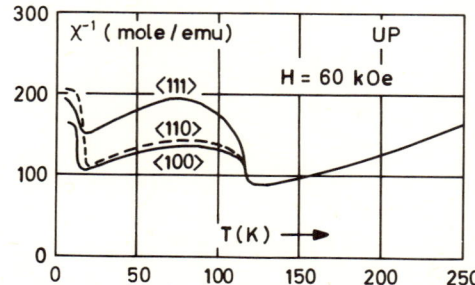

Fig. 8. Inverse magnetic susceptibility χ^{-1} vs. temperature T of a single crystal of UP measured in a field of $H = 60$ kOe, applied in three different crystallographic directions. From Ref. 28. Later measurements (Ref. 258) find that $\chi^{-1}(T=0)$ is 500 mole/emu in the $\langle 110 \rangle$ direction; see Fig. 9.

2.2. NaCl-TYPE METALLIC ACTINIDE COMPOUNDS

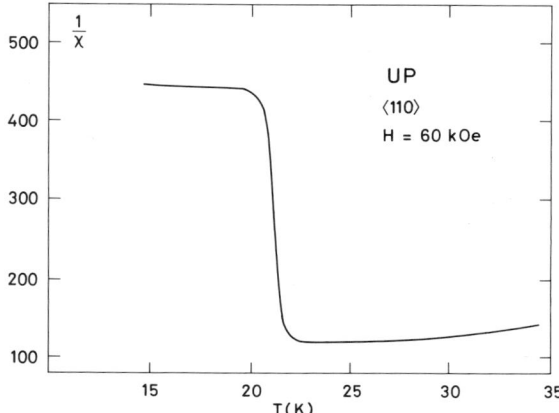

Fig. 9. Inverse susceptibility of UP along the ⟨110⟩ axis vs. temperature T in magnetic fields of different strengths. From Ref. 258.

Data on arcmelted samples on UP show that the moment-jump transition at $T = T'$ is accompanied by a sudden increase in the electrical resistivity[6,7] ρ (see Fig. 11). Between T' and T_N there is an approximately linear increase in the resistivity, which becomes almost temperature-independent between T_N and 300 K.[6] The lattice constant a of the cubic cell of UP also varies anomalously as

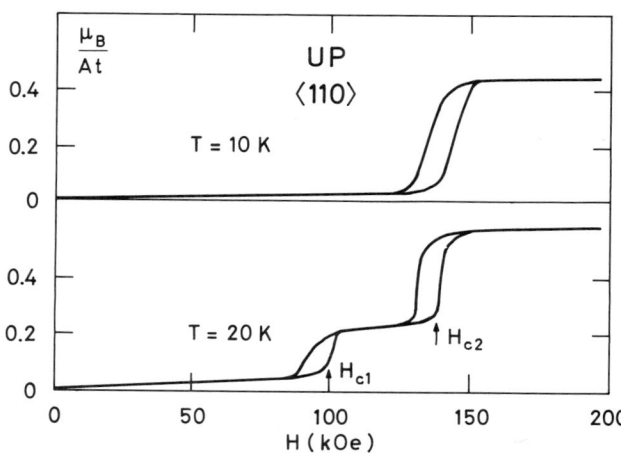

Fig. 10. Magnetization of UP vs. field applied along the ⟨110⟩ axis for different temperatures. H_{c_1} and H_{c_2} are critical fields for inducing transitions to new, yet unknown spin structures. From Ref. 257.

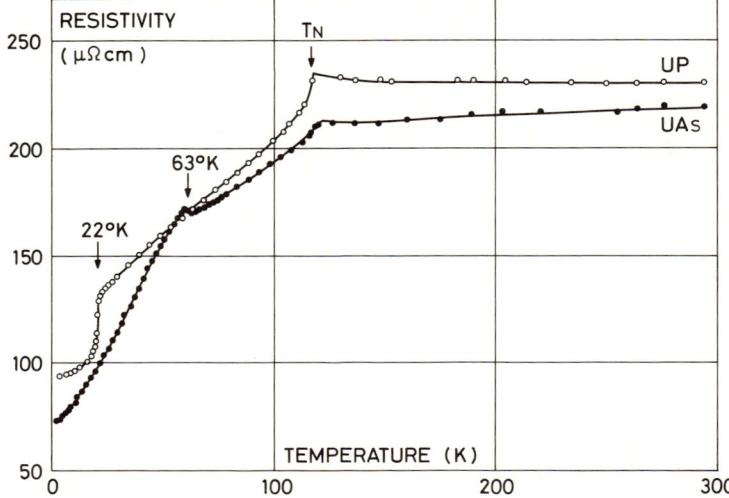

Fig. 11. Electrical resistivity vs. temperature T of UP and UAs. Notice the small peaks at the antiferromagnetic phase transition temperatures T_N and the anomalies at the temperatures indicated by arrows, where additional phase transitions occur. From Ref. 6.

a function of T near $T = T'$ (see Fig. 12).[9] The most precise data reveal small discontinuities in the volume V both at the moment-jump transition ($\Delta V/V \cong 2 \times 10^{-4}$) and at T_N ($\Delta V/V \cong 3.6 \times 10^{-5}$),[253] thus confirming the first-order nature of both transitions.

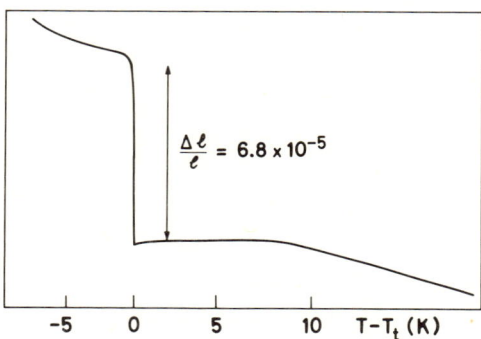

Fig. 12. Change, Δl, in the length l of a single crystal of UP at the transition temperature $T_t = 22.5$ K, from Ref. 253. From this, a corresponding change of the lattice constant may be deduced.

2.2. NaCl-TYPE METALLIC ACTINIDE COMPOUNDS

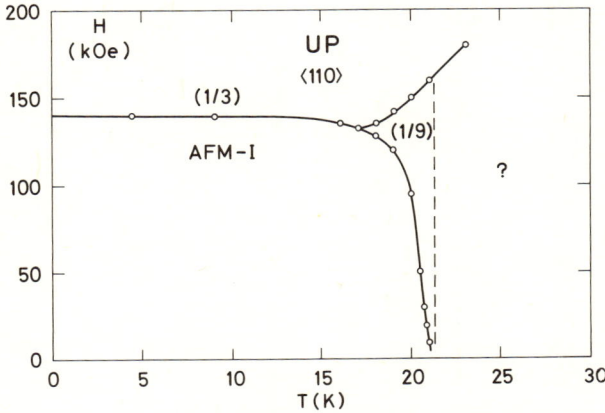

Fig. 13. Magnetic phase diagram of UP for field applied along the ⟨110⟩ axis. The notations "1/3" and "1/9" are symbols for the unknown ferromagnetic spin structures, with ferromagnetic components approximately equal to 1/3 and 1/9 of the saturation moment, respectively. From Ref. 257.

The behavior of UP seems unique in the field of magnetism. Discontinuous changes in the ordered moment as a function of T are observed for some rare earths, but they are associated with changes in the magnetic ordering, e.g., transitions from a flat spiral to a cone.[59] Upon dilution of UP with USe, temperature-induced, moment-jump transitions persist up to a USe concentration of 20% (see Fig. 47, p. 49).[260] The experimenters claim that the U moments are aligned in the ⟨001⟩ direction at temperatures below and above the transition, but again it should be cautioned that the samples were not single crystals. A magnetic phase diagram proposed for single crystal UP in the ⟨110⟩ direction is shown in Fig. 13.[257]

Solid solutions of UP and ThP have been prepared over the whole range of 0-100% of uranium. The results[252] of measurements of the lattice constant a, of the Curie-Weiss constant θ, and of the effective magnetic moment of the U ion are shown in Figs. 77c and 78c (p. 80). It is not clear whether the deviation from a straight line of either one of these three experimental curves is significant in view of the experimental uncertainties. Indeed, the latest data of Troć[261] show a linear dependence of a vs. x. It is interesting to compare these results with the same quantities measured on UO_2-ThO_2 solid solutions (cf. Section 2.6.2 and Fig. 78). It may be seen that the magnetic moment varies in both sets from $\simeq 3.3\,\mu_B$ for the concentrated to $2.8\,\mu_B$ for the dilute uranium (discussed in the Section on crystal-field theory) and that the Curie-Weiss constant does not extrapolate to zero uranium concentration (discussed in Section 2.6.2).

2.2.3. Uranium Monoarsenide

Uranium monoarsenide, **UAs**, is very similar to UP in its electrical and magnetic properties, as Table 3 indicates. The sublattice magnetization from 4.2 K to 90 K has been determined from several independent neutron diffraction experiments[2,3,60,61] and the results are shown in Fig. 14a. At $T = 4.2$ K, the type of magnetic ordering is the "two k" form of AFM-IA (see Fig. 5).[254,262] In this structure the ordered U-moments lie in equivalent $\langle 110 \rangle$ directions, the signs of their x and y components alternating in the sequence (++−−) characteristic of type IA ordering. Upon heating to $T' = 63 \pm 2$ K, the ordered moment μ drops suddenly from $\mu = 2.24\,\mu_B$ to $1.92\,\mu_B$ per uranium ion, and the magnetic ordering transforms to AFM-I (Fig. 4b) with the moments aligned in the $\langle 001 \rangle$ direction. Experimentally, a transition width of about 7 K was found in which the neutron diffraction lines of both the AFM-IA and AFM-I phases could be seen.[2] One may expect that more refined experiments will reduce this transition width, since in the case of UP the first neutron diffraction results[55] also indicated a broad transition around $T' \simeq 22.5$ K; whereas later the more accurate NMR results[1] revealed a width of only 0.2 K for the moment-jump transition. No NMR experiments have yet been reported on UAs. The first-order nature of the transition at $T' = 63$ K is also indicated by the absence of critical scattering[2] and the steep jump in the magnetic powder susceptibility[6] (Fig. 14b) and a kink in electrical resistivity[6] (Fig. 11).

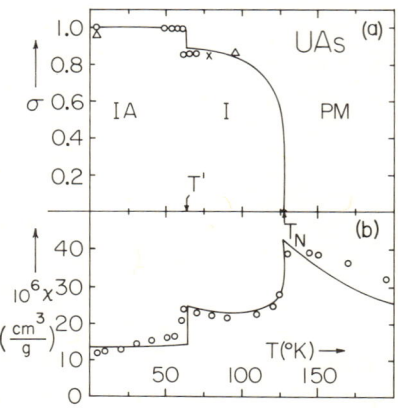

Fig. 14. a). Sublattice magnetization σ of uranium arsenide UAs *vs.* temperature T. Circles, triangles, and cross: Experimental results of Refs. 2, 3, and 3, respectively. b). Powder susceptibility χ *vs.* temperature T. Circles: Experiments of Ref. 6., IA, I: Antiferromagnetic ordering types (see Fig. 4). PM = Paramagnetism. Solid lines in a) and b): Theory of Ref. 16.

Another anomalous feature of the sublattice magnetization curve of UAs is that the ordered moment shows no detectable decrease in the intervals $4.2\,K < T < T'$ and $T' < T < 96\,K$.[2] The "flatness" of the σ vs. T curves above and below $T = T'$ indicates that elementary magnetic excitations of the effective field or spin wave types are practically absent, because they would cause a smooth decrease of the ordered moment with temperature in each phase. The single-crystal neutron data reveal that the transition at $T = T_N$ is strongly first-order,[262] because the intensity of the [201] magnetic peak vanishes quite suddenly at the latter temperature. This is also indicated by the sharp jump in the susceptibility (Fig. 14b) at $T = T_N$. Neutron diffraction data on the related compound $UAs_{0.68}S_{0.32}$[2] also show a rapid drop of the magnetic moment at T_N.

Measurements of the magnetic moment of single crystals of UAs show that at 4.2 K fields of 70 kOe create a single-domain crystal[28,62] (Fig. 15). This gives an indication of the magnitude of the magnetic anisotropy field. In lower fields and powder samples hysteresis effects are observed due to domain structure.[58]

As in the case of UP, fields of $\cong 150\,kG$ cause transitions to structures with net ferromagnetic components,[28] with the easy axis lying in the $\langle 110 \rangle$ direction in agreement with the latest diffraction results.[262]

In applied magnetic fields of $\cong 45\,kOe$ or greater, the specific heat of UAs as a function of temperature develops a double-peaked structure near T_N[262] (Fig. 16). The lower peak (near 121 K for $H = 70\,kOe$) represents a first-order transition from the AFM-I to a ferrimagnetic state, and the second peak (near 127 K for $H = 70\,kOe$) results from a transition from the ferrimagnetic state to paramagnetism. An applied external pressure (approximately 8 kbar for $H = 70\,kOe$) was found to eliminate the ferrimagnetic phase.[262] The effect of pressure is shown in Fig. 17. The magnetic behavior of UAs is summarized in the phase diagram[446] shown in Fig. 15b.

The resistivity as a function of temperature shows the same qualitative features as in UP. Two kinks at T' and T_N, respectively, with a roughly linear increase in the intervals 4.2–63 K and 63–127 K[6] are observed (Fig. 11).

Inelastic neutron scattering experiments[412] on UAs found a broad magnetic contribution with a half-width varying from $8 \pm 1\,meV$ at 5 K to $15 \pm 2\,meV$ at 250 K. Although the presence of acoustic and optical phonons could be inferred from the data, well-defined collective (e.g., spin wave) excitations could not be seen.

The similarities between the data for UP and UAs suggest that the same physical model should explain the transitions in both compounds, even though the moment jump in UAs occurs at a much higher temperature than that of UP

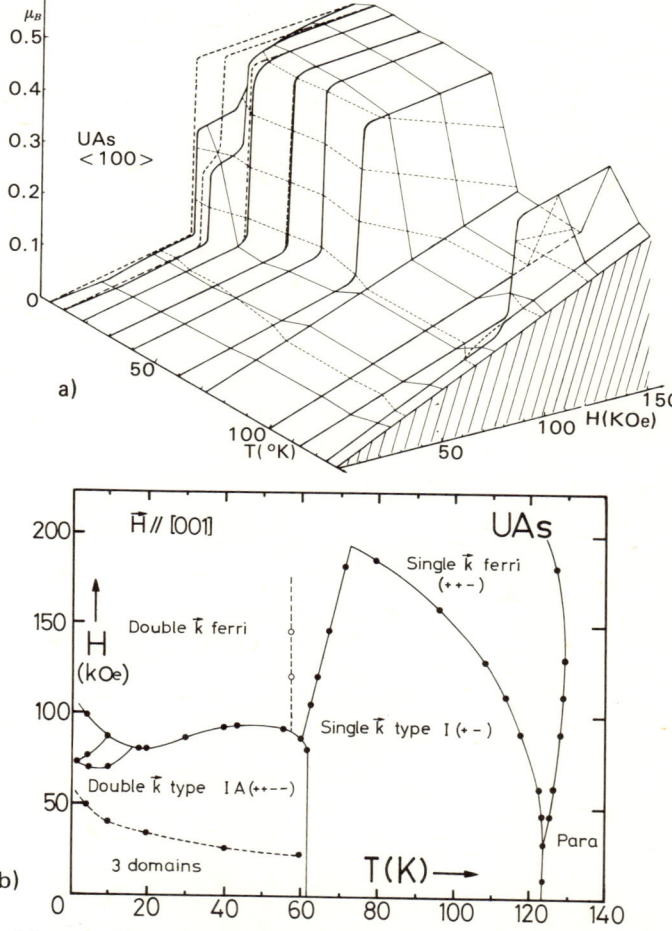

Fig. 15. a). Magnetic moment per U ion of a single crystal of UAs, measured in the ⟨100⟩ direction, as a function of both the temperature T and the applied magnetic field H. Notice that in the low-temperature region a step-behavior is observed which the authors tentatively interpret as due to the formation of intermediate spin structures. From Ref. 62.
b). Magnetic phase diagram in the H, T plane of a UAs single crystal in a magnetic field applied along the [001] crystallographic axis. The diagram is constructed on the basis of neutron diffraction data. In the double **k** ferrimagnetic phase, $k = 0.66$, and in the k_{xz} domains $\mathbf{k}_1 = [\frac{1}{2} 0 0]$, $\mathbf{k}_3 = [00k]$; in the k_{yz} domains $\mathbf{k}_2 = [0\frac{1}{2}0]$ and $\mathbf{k}_3 = [00k]$. In the single **k** ferrimagnetic phase k_z domains, $\mathbf{k}_3 = [00\,2/3]$. In the single **k** type-I phase k_x domains, $\mathbf{k}_1 = [100]$ and in the k_y domains $\mathbf{k}_2 = [010]$. In the double **k** type-IA phase k_{xy} domain $\mathbf{k}_1 = [\frac{1}{2}00]$ and $\mathbf{k}_2 = [0\frac{1}{2}0]$. In the three-domain phase these are of k_{xy}, k_{yz}, and k_{zx}-type. From Ref. 447.

2.2. NaCl-TYPE METALLIC ACTINIDE COMPOUNDS

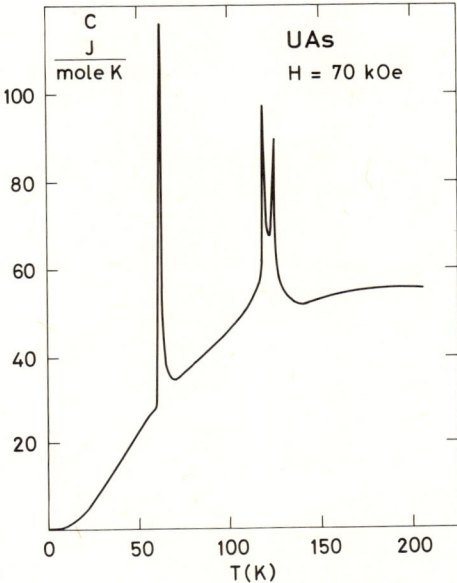

Fig. 16. Variation of the specific heat of UAs for a constant magnetic field applied along the ⟨001⟩ axis, as a function of the temperature T. From Ref. 262.

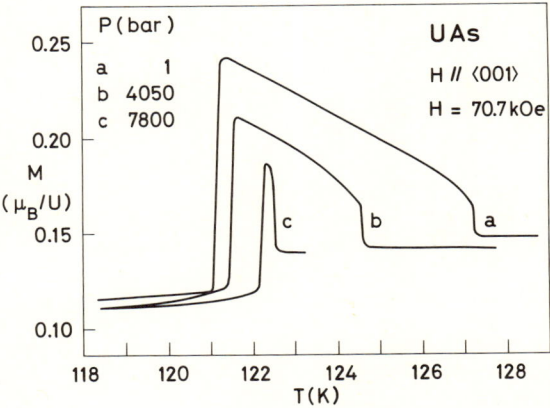

Fig. 17. Pressure effects on the thermal variation of the magnetization for a field of 70.7 kOe applied along the ⟨001⟩ axis of a UAs single crystal. From Ref. 262.

2.2.4. Uranium Mononitride

Uranium mononitride, **UN**, is a Type I antiferromagnet[24,62] like UP and UAs with a Néel temperature $T_N = 53.1 \pm 0.2$ K,[64] but differs from the latter two in having a markedly smaller lattice constant ($a_0 = 4.89$ Å)[64] and a smaller value $\mu_0 = 0.75 \mu_B$ of the ordered moment at $T = 0$ K. The sublattice magnetization curve obtained from neutron diffraction[63] and shown in Fig. 18 reveals no evidence of a moment-jump transition, but the data do not extend below 12 K. A very interesting feature is the comparatively small peak at $T = T_N$ in the specific heat *vs.* T data of Counsell *et al.*[46] (see Fig. 19). The entropy increase associated with the magnetic disordering at T_N as deduced from the specific heat data is only 0.15 cal deg^{-1} mole^{-1}. This may be compared with the values of about 1 cal deg^{-1} mole^{-1} found for other uranium compounds. It was suggested[46] that the entropy of disorder of the magnetic moments is partially compensated by an entropy decrease caused by a redistribution of electrons among band states and that, in fact, UN is a band or itinerant antiferromagnet such as Cr metal. Further evidence for this suggestion is provided by experiments showing that T_N and μ_0 decrease rapidly under pressure in a manner expected for a band antiferromagnet (Fig. 20).[263]

There is certainly an extremely high band electronic heat capacity coefficient γ (35 times that of copper), (see Table 3), which points to a high density

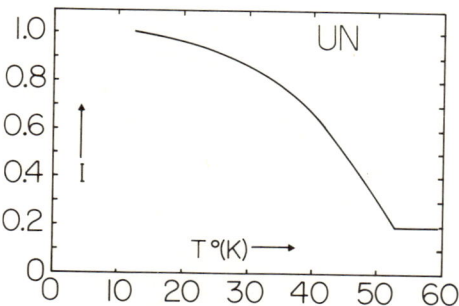

Fig. 18. Neutron counting rate, I, in diffraction experiment *vs.* temperature T of uranium nitride, Ref. 63. The ordinate is a measure of the square of the ordered moment.

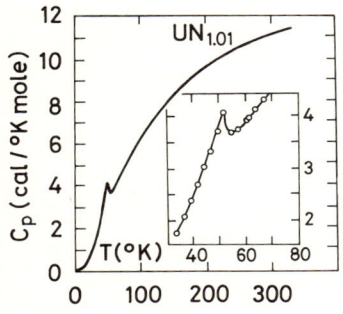

Fig. 19. Specific heat at constant pressure, c_p, vs. T of UN, Ref. 46. The insert shows details near T_N.

of band electronic states at the Fermi energy and would support the idea of band antiferromagnetism.

On the other hand, refined specific heat measurements from 1.3 to 4.6 K yield[65] a Debye temperature of 324 K, which, when used to estimate the magnetic entropy, yields much larger values than the 0.15 cal deg^{-1} mole^{-1} deduced previously from $\theta_d = 276$ K.[46] This removes the necessity to seek an explanation for an abnormally small entropy of disorder. A final point of interest is the curvature in the inverse paramagnetic susceptibility[66] $\chi^{-1}(T)$ (Figs. 21, 22), which the Curie–Weiss law would predict to be linear.

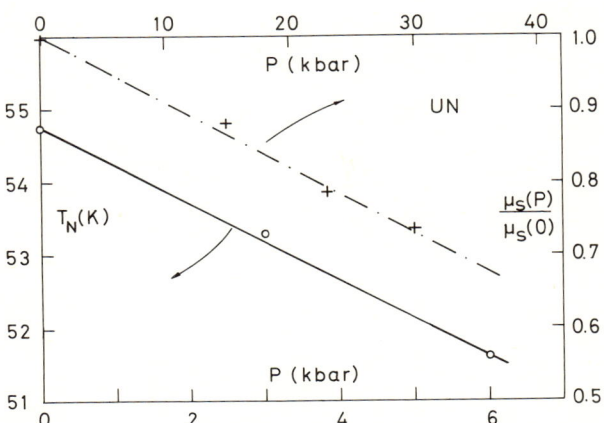

Fig. 20. Pressure dependence of the Néel temperature (bottom and left scales, full line) and pressure dependence of the saturation moment (top and right scale, dash-dotted line) of UN. From Refs. 388 and 263, respectively. The authors note that $d \ln T_N/d \ln V = d \ln \mu_s/d \ln V = 19$, where $V =$ volume.

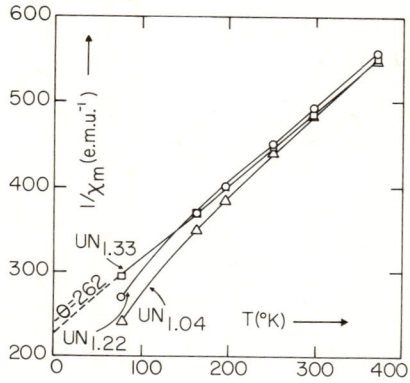

Fig. 21. Inverse molar susceptibility χ^{-1} vs. T of UN_x, from ref. 66. The dashed curve is an extrapolation from which the Curie–Weiss constant θ is deduced.

A "modified Curie–Weiss law," with $\chi = \chi_0 + C/(T-\theta)$, however, yields a straight line for $(\chi - \chi_0)^{-1}$, and leads to the lower value of θ and higher value of μ_s in Table 3. χ_0 is found to be 8.0×10^{-6} emu cm^{-3}.[67]

Electrical resistivity measurements between 4 K and T_N on single crystals are fairly well described by the empirical relationship $\rho = [1.18 + 77.4(1 - M^2)]\mu\Omega$cm, where M is the measured sublattice magnetization (Fig. 23). At T_N a small peak is observed and from T_N to 300 K the resistivity increases by a factor of two. The resistivity behavior below T_N is attributed to spin-disorder scattering, the peak to the formation of new Brillouin-zone boundaries upon antiferromagnetic ordering.

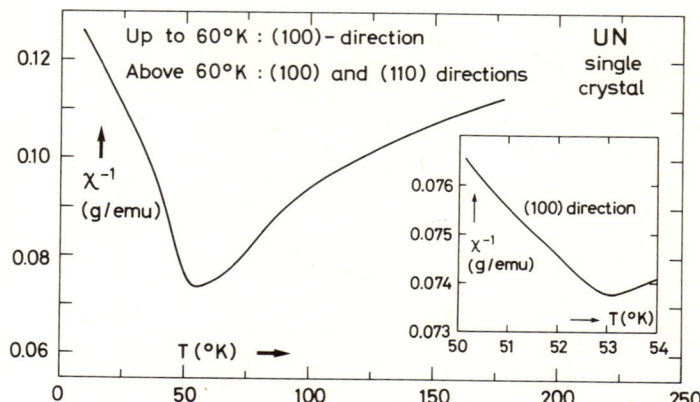

Fig. 22. Inverse gram magnetic-susceptibility of single crystal UN vs. T. Insert shows details near T_N. From Ref. 67.

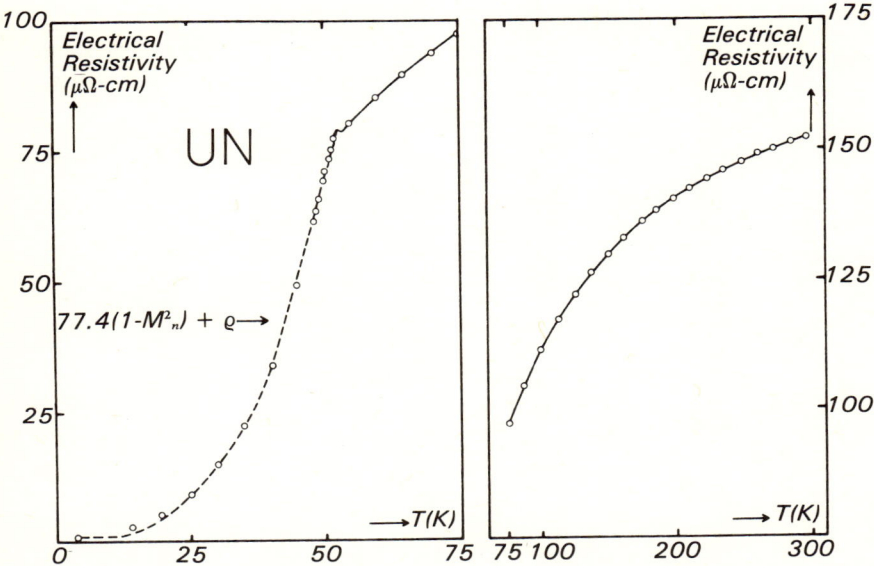

Fig. 23. Electrical resistivity (circles and solid line) vs. temperature T of a single crystal of UN along the $\langle 100 \rangle$ direction. The broken line is a fit with the empirical formula shown on the diagram where M_n is the sublattice magnetization. From Ref. 67.

In the magnetically ordered phase a tetragonal distortion has been observed.[68] The corresponding strain measured in a multidomain crystal along the $\langle 100 \rangle$ direction is exactly proportional to the square of the sublattice magnetization, and reaches a maximum value of about $\Delta l/l = 2 \times 10^{-4}$.[68] A similar behavior is found in the elastic constant C_{44} (Fig. 24).[24]

Photoelectron spectroscopy experiments on UN revealed a narrow band of f states near the Fermi level containing $2.2 \pm .5$ electrons.[264] This indicates the possibility of intermediate valence for uranium in UN – that is, the number of electrons in the $5f$ shell may not be integral.

The optical phonon spectrum of UN and of $U(C_{0.6}N_{0.4})$ has been measured[50] by the method of high-resolution (time-of-flight) inelastic neutron scattering. The dispersion curves can be well fitted by only two short-range radial force constants, one between the actinide and X (X = C, N), and another between nearest actinide atoms. In the mixed compounds, the optical phonon peak is split in two, corresponding to local vibrations of C and N atoms. Neutron measurements also reveal that the moments are aligned in the $\langle 001 \rangle$ direction and couple much more strongly to moments within the (001) planes than to moments in neighboring planes.[265,71] The magnetic excitation spectrum is

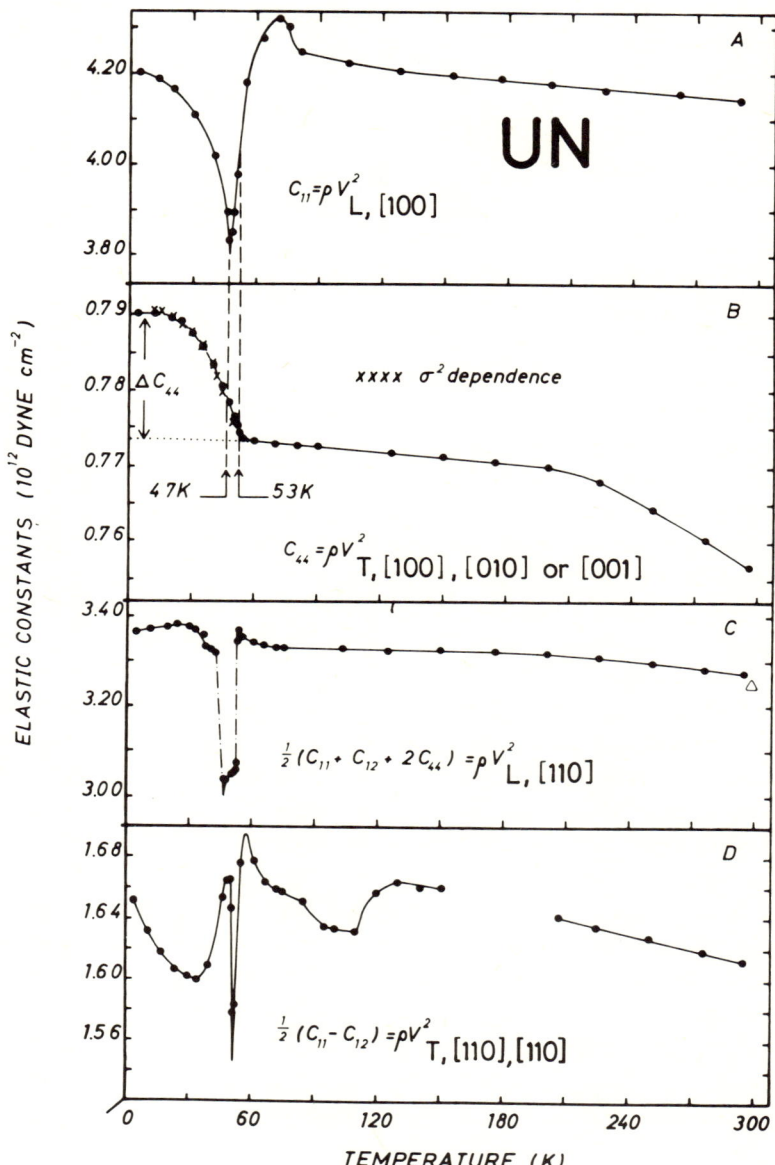

Fig. 24. *A, B, C, D*: different combinations of elastic constants c_{ik} as measured by the transverse (*T*) or longitudinal (*L*) sound velocities *V*, in certain crystallographic directions (subscripts [*klm*]) *vs.* temperature *T* for UN monocrystals. Crosses in B: square of the sublattice magnetization (arbitrary scale). From Ref. 24.

longitudinal, has a large anisotropy gap, and cannot be easily explained by conventional spin waves.

2.2.5. Uranium Antimonide

Uranium Antimonide, USb, is a type I antiferromagnet with a Néel temperature of $T_N = 213\,\text{K}$.[267] Crystals were thought to contain three domains, but newer data[259,266] point to a *3k* structure (Fig. 48). Neutron diffraction[69] yields values of the atomic moment between $2.16\,\mu_B$ and $2.85\,\mu_B$, whereas from paramagnetic susceptibility measurements on single crystals $\mu_p = 3.64\,\mu_B$ is deduced.[28] The inverse susceptibility between T_N and $1100\,\text{K}$, measured in the $\langle 100 \rangle$ direction, is a straight line ($\theta = 140\,\text{K}$) in contrast to UP and UAs which show a slight curvature.[28] At T_N χ^{-1} exhibits a minimum just as in UP and UAs. The magnetization at $1.3\,\text{K}$ as a function of the applied field of a powdered sample is linear, and reaches $\cong 0.15\,\mu_B$ per U atom at 30 Tesla.

Optical reflectivity measurements on single crystals were performed in the photon range of $0.03-11\,\text{eV}$.[266] With the help of the Kramers–Kronig relations the authors deduce the complex dielectric function: in particular they identify the transitions $p(\text{Sb}) \rightarrow 6dt_{2g}$ and $p(\text{Sb}) \rightarrow 6de_g$. A definite interpretation of the spectrum will have to wait until more information regarding the band structure and optical transition matrix elements becomes available on USb, but an intermediate valence and/or hybridization of the *f* and *p* states were suggested.[266]

An intriguing observation is that[70] of longitudinal collective (presumably magnetic) excitations between $8-100\,\text{K}$ (Fig. 25). This contrasts with the usual transverse spin wave modes. The authors point out the inadequacy of present theoretical models to understand these spin waves. We note that similar excitations have been observed in UN.[71]

Jensen and Bak[292] have shown that the inelastic neutron scattering data on USb can be well explained as spin-wave excitations in the triple-magnetic structure proposed by Rossat-Mignod et al.[262] The former authors take into account the crystal field, the anisotropic exchange, and an anisotropic interaction of the dipolar form

$$H = - \sum_{i \neq j} J_D (\mathbf{S}_i \cdot \mathbf{r}_{ij})(\mathbf{S}_j \cdot \mathbf{r}_{ij}),$$

where the sum is restricted to nearest neighbors connected by vectors $\mathbf{r}_{ij}$. In their model, the triple-*k* structure (see Appendix B) in which the actinide spins point in $\langle 111 \rangle$ directions is favored by the crystal field. Carrying out a standard spin-wave calculation, Jensen and Bak find a low-lying longitudinal mode and a

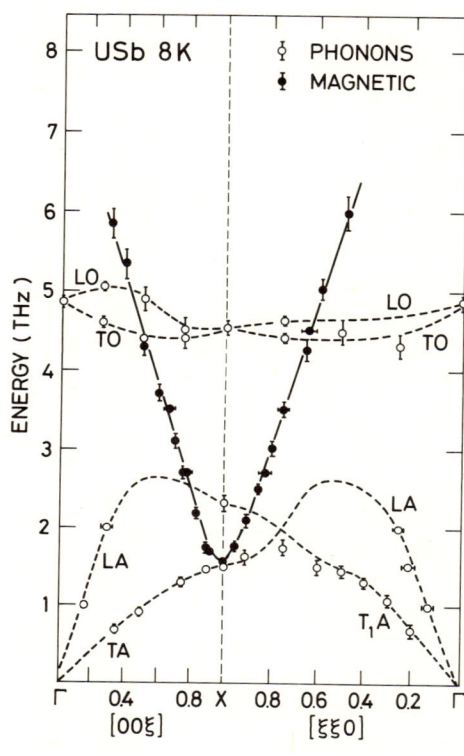
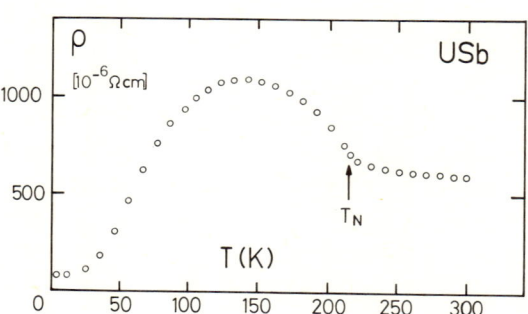

Fig. 25. Top). Dispersion curves of excitations observed by inelastic neutron scattering in an assembly of six oriented single crystals of USb at $T = 8$ K. The frequency of the excitation is plotted vs. wave-number transfer (in units of $2\pi/a$, a = lattice constant) in two directions. The magnetic modes are represented by the solid line fitted to the measurements (solid dots), the phonons by dashed lines drawn on the basis of the measurements (open circles) and the knowledge of phonons in NaCl-structures. Not shown is a broad non-dispersive, high-energy magnetic mode (5–8 THz). From Ref. 70. Bottom) Temperature dependence of the electrical resistivity of a USb single crystal. There is a remarkable absence of drastic change in ρ at the Néel temperature T_N. The broad peak below T_N is attributed to the possible appearance of energy gaps in the electronic band structure at the Fermi energy due to the magnetic Brillouin zone boundaries which develop below T_N. From Ref. 445.

higher, flat, transverse mode. For reasonable values of the three model parameters (i.e., the isotropic and anisotropic exchange and the B_4^0 crystal-field parameter), the experimental excitation spectrum in the [00ξ] and [ξξ0] directions in the Brillouin zone is quantitatively explained. Because the model assumed semiclassical spins, it is independent of the detailed ground state of the 5f electrons, and by the same token provides no information about that state.

The electrical resistivity ρ of USb as a function of temperature T (Fig. 25, bottom) shows two interesting features.[445] First, the anomaly in $\rho(T)$ at the ordering temperature T_N is very small; and second, there is a very broad maximum in $\rho(T)$ occurring at $T \cong 150\,\text{K}$. It has been suggested[445] that the latter maximum results from a modification of the conduction electron Fermi surface by the superlattice of the 3k magnetic structure proposed by Rossat-Mignod et al.[262]

2.2.6. Uranium Monchalcogenides

The *uranium monochalcogenides* **US**, **USe**, and **UTe**, are ferromagnetic semimetals with ordering temperatures of $T_C = 180\,\text{K}$, $160\,\text{K}$, and $108\,\text{K}$, respectively.[72,73] The values μ_0 of the ordered moments at $T = 0\,\text{K}$ as deduced from neutron diffraction[55] are listed in Table 3 and are considerably higher than the values $\mu_0 = 1.1\,\mu_B$, $1.31\,\mu_B$, and $1.1\,\mu_B$ derived earlier from bulk magnetization measurements[19,74,75] on UTe, US, and USe, respectively. In these materials, there exists a tremendous magnetic anisotropy field,[76] larger than $10^6\,\text{G}$ in US,[72] which makes the $\langle 111 \rangle$ direction the "easy axis" along which the uranium magnetic moments are directed.[258,253] There is also a rhombohedral distortion which sets in below T_C in these compounds.[77,78] In US, the rhombohedral angle is $89.61°$ at $4.2\,\text{K}$. The variation of the lattice constant and the rhombohedral angle of UTe with T are shown in Fig. 26a and b.[79,268] One possible source of the small values of μ_0 found by the magnetization measurements is a conduction electron spin polarization induced by the applied field

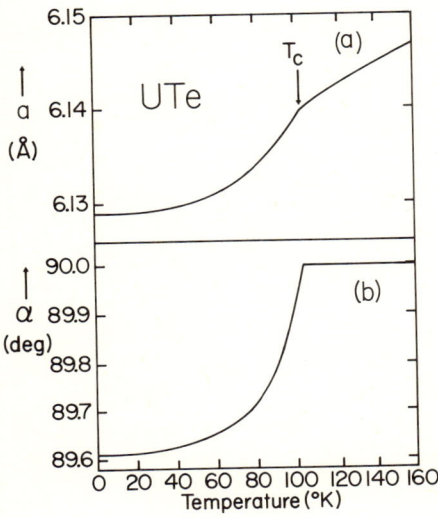

Fig. 26. Uranium telluride, UTe: a). Lattice constant a vs. temperature T, Ref. 79. b). Interaxial angle α vs. T. $(90° - \alpha)$ represents a deformation from cubic to rhombohedral structure. From Ref. 79.

with the conduction electron spins antiparallel to the local moments.[80] In zero applied field however, the neutron diffraction results for US indicate[72] that the spin polarization is uncorrelated with the localized moment.

Magnetization *vs.* temperature data in a 1 kG applied field[74] show a normal Brillouin-type curve for US as well as for UTe in a 10 kG field,[258] but the behavior of USe reveals a maximum in the magnetization below T_C (see Fig. 27). The origin of this maximum in USe is as yet unknown. The inverse paramagnetic susceptibility[74] $\chi^{-1}(T)$ of US and USe deviate strongly from the Curie–Weiss law above T_C, as seen from Fig. 28.

The measurement of the hyperfine field at the Te nucleus by Mössbauer spectroscopy as a function of temperature[81] yields a curve in complete agreement with that of the ordered U-moment obtained from neutron spectroscopy.[73]

The high values of $\gamma = 56 \times 10^{-4}$ cal/mole-deg^2 and $\gamma = 208 \times 10^{-4}$ cal/mole-deg^2 found[82,83] for the coefficient of the linear term in the low-temperature specific heat of US and USe, respectively, indicate a narrow electronic band at the Fermi level. Photoemission measurements[84,269,270] are also evidence that the 5f energy levels form a narrow band intersected by the Fermi surface. The temperature dependence of the galvanomagnetic properties of US have been explained[34,42] as resulting from thermal redistribution of electrons among overlapping bands. The effective carrier concentration was found to be 0.45 holes per U atom.[42] This is in rough agreement with band calculations which predict a transfer of 0.6 uranium f electrons to the valence band.[271]

These predictions of a fractional (or mixed) U valence also compare favorably with a recent estimate[414] of a fractional valence of 3.5 in US. The estimate is based on an empirical scheme in which the uranium valence is

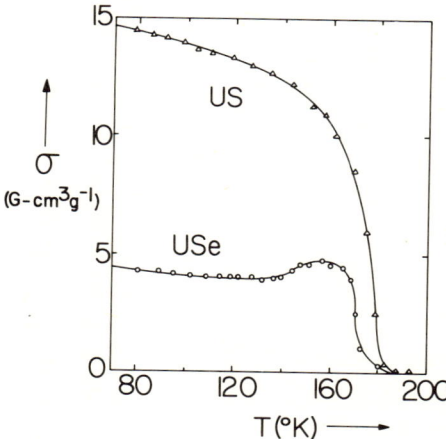

Fig. 27. Magnetization σ, in an applied field of 1 kG, of uranium selenide USe and sulfide US, as functions of the temperature T. From Ref. 74.

2.2. NaCl-TYPE METALLIC ACTINIDE COMPOUNDS

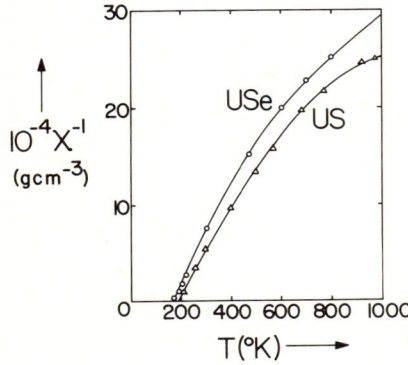

Fig. 28. Inverse susceptibility χ^{-1} of USe and US, as functions of the temperature T. From Ref. 74.

deduced from the distance between the uranium and the chalcogen atom. The same scheme yields fractional valences of 3.4 and 3.2 in USe and UTe, respectively. The assumption of a correlation between the U-chalcogen distance and valence is, however, more valid for rare earth than for actinide compounds because of the more complicated nature of the f electronic states of the latter compounds.

A negative (i.e., opposite to that of U) spin polarization of photo-emitted electrons has been observed[85] in UTe[85,86] and in US[87] (Fig. 29) throughout a spectrum of 7 eV below the Fermi level. This is shown theoretically[87,88] to arise from the negative spin polarization of the $6d$ bands and not $5f$ electrons at or close to the Fermi level, which have a vanishing photoemission probability. The $6d$ band is split not only into the t_{2g} and e_g parts, but due to the f–d interaction, there occurs also an asymmetric splitting of the t_{2g} subband. This creates a trough in the density of states of the positive-spin d electrons at and below the Fermi surface, a feature also seen in the energy-level scheme deduced from optical spectroscopy.[272]

If one assumes, as mentioned earlier in this section, that the bulk magnetization is due to the sum of f- and d-polarizations, whereas the neutron scattering experiment measures only the former, localized moment, the above-mentioned theory may also explain why the bulk magnetization per U ion is only half as big as the one found by neutron scattering.

A new phase (US III) of US was recently observed to form under external applied pressures greater than 15 GPa.[413] This phase appears to be an orthorhombic distortion of a tetragonal lattice with parameters $a = 3.64$ Å, $b = 3.66$ Å, and $c = 5.63$ Å. The experiments are continuing with the aims of bringing about the disappearance of magnetism and delocalizing the $5f$ electrons by application of higher pressures, i.e., achieving a Mott transition.

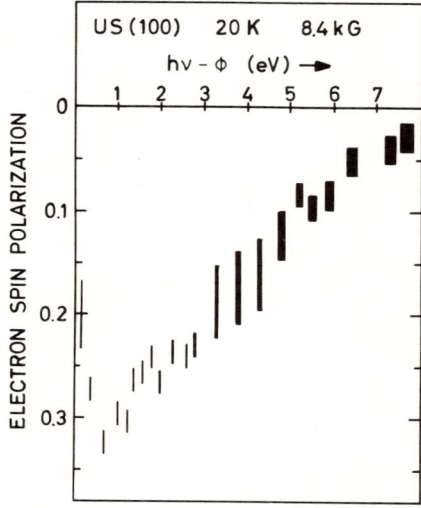

Fig. 29. Spin polarization of electrons ejected from a (100)-crystal face of ferromagnetically-ordered US cleaved in ultra-high vacuum. The abscissa is the photon energy $h\nu$ minus photo threshold Φ. The negative sign of the ordinate (1 corresponds to 100% polarization) indicates that the photoelectrons are polarized opposite to the applied field of 8.4 kG, which is also the polarization direction of the U-ions. Experiment at temperature $T = 20$ K. From Ref. 87.

The magnetic and phonon excitations in UTe have been determined by inelastic neutron scattering at 4.2 K.[265] There appear to be five spin-wave branches with a large (3.45 THz) anisotropy gap and considerable intrinsic breadth due to scattering from unknown causes.

The measurements of the temperature dependence of resistivity[415] in single crystals of USe revealed a term of the form $T^2 \exp(-\Delta/T)$, where $\Delta = 180$ K was deduced from fitting the data. The quantity Δ was interpreted as a gap at the wave vector $q = 0$ in the spin wave spectrum, not yet observed by inelastic neutron scattering. The same procedure applied to US yields[415] $\Delta = 0$.

Du Plessis et al.[448] determined the critical exponent $\beta = 0.24 \pm 0.01$ for USe by a neutron scattering experiment in the expression $M \propto (T/T_c - 1)^\beta$. Previously Aldred et al.[449] obtained $\beta \cong 0.29$ for UTe, and van Doorn[450] finds $\beta \cong 0.25$ for US. These data contradict theoretical expectations of $\beta \cong 0.345$ for the Heisenberg model, but also those for a 3-d Ising model of $\beta \cong 0.313$.

On the other hand, bulk magnetization studies[449] gave $\gamma = 1.33$ for UTe as predicted by the Heisenberg model.

2.2.7. Neptunium Carbide

This material has been prepared only in the form NpC_x, where values of x from $x = 0.82$ to $x = 0.96$ have been reported.[7] The missing carbon atoms yield a random distribution of vacancies in the NaCl lattice. Neutron diffraction

2.2. NaCl-TYPE METALLIC ACTINIDE COMPOUNDS

experiments[4] on a powder sample of $NpC_{0.93}$ show that it is a ferromagnet below $T' = T_C = 220$ K with the moments aligned along a cubic axis. Upon additional heating the ordered moment σ per Np atom drops from $(1.6 \pm 0.1)\mu_B$ to $(1.3 \pm 0.1)\mu_B$, and the moments of every second layer turn over, i.e., there is a transition to AFM-I ordering. The magnetization vs. T curve above T' does not follow a Brillouin curve as Fig. 30a shows. There appears to be an inflection point in the σ vs. T curve near $T = 260$ K which may correspond to the peak in the specific heat observed by Sandenaw et al.[89] at the latter temperature (see Fig. 30c). The specific heat data were not reproducible in the range 240–310 K, and the neutron diffraction results were highly inaccurate above $T = 280$ K.[4] Thus, the nature of the magnetic transitions above $T = T'$ is still unclear. A small peak in the specific heat at $T = 310$ K can be seen, corresponding to the Néel temperature T_N found by neutron diffraction.

In the FM phase, there is a rhombohedral distortion in NpC similar to that observed in US.[78] The peculiar thing is that the distortion in NpC occurs not at $T_C = 220$ K but at a temperature about 6 K *below* T_C.

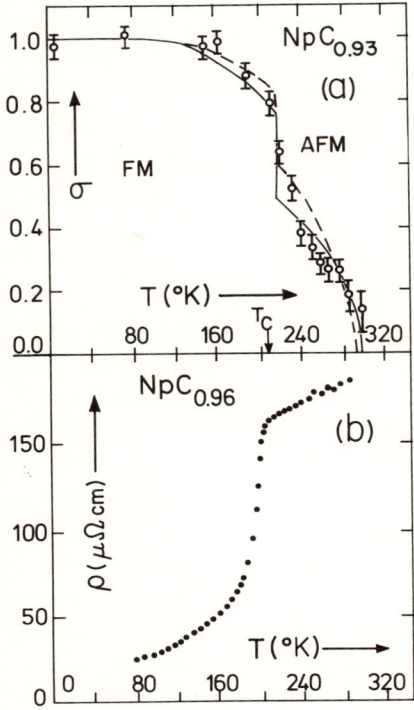
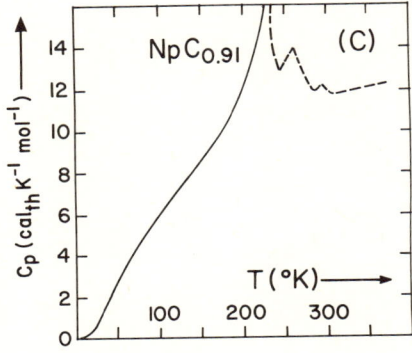

Fig. 30. Neptunium carbide, NpC: a). Relative ordered moment σ vs. temperature T, Ref. 4. b). Resistivity ρ vs. T, Ref. 7. (The curves are results of a theory discussed in Chapter 3.) c). Specific heat at constant pressure c_p vs. T, Ref. 89. The dashed curve shows uncertain data.

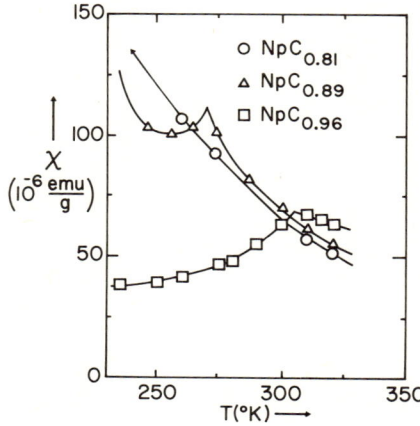

Fig. 31. Magnetic susceptibility χ of neptunium carbide as a function of temperature T. With decreasing carbon concentration the position of the peak which indicates the Néel temperature shifts to a lower temperature. From Ref. 7.

A most striking feature of the behavior of NpC is the steep increase in the electrical resistivity[7] ρ as T is raised through the moment-jump transition temperature $T' = 220$ K (see Fig. 30b). The ρ vs. T curve is similar to that of UP (Fig. 11a), but the size of the jump is larger in NpC.

In Fig. 31, the magnetic powder susceptibility[7] χ as a function of T is plotted for several NpC_x samples. One sees that T_N varies strongly with x but not T' and that there is no indication of a third transition between T' and T_N as the three peaks in c_p vs. T (Fig. 30c) would suggest. The Néel temperature of 310 K found[4] by neutron diffraction for the sample $NpC_{0.93}$ may be too high because the susceptibility data show[7] $T_N \cong 280$ K for $NpC_{0.91}$.

2.2.8. Neptunium Monopnictides

In addition to NpC, discussed above, there are several other neptunium compounds that show very unusual magnetic phase transitions.[40,78,90] These are the *pnictides*, i.e., compounds formed with the VA group elements N, P, As, Sb, and Bi of the periodic table. Neptunium phosphide, **NpP**, has been investigated[40,78] by neutron diffraction and found to be antiferromagnetic below $T_N = 130$ K. The type of magnetic ordering is not one of the types previously discussed, however. The magnitudes and signs of the ordered moments on the Np sites vary sinusoidally in the lattice, giving rise to a longitudinal magnetization wave. This wave is incommensurate with the lattice spacing above 74 K, but below 74 K is commensurate with a period of three unit cells. At $T' = 74$ K there occurs a moment-jump transition similar to that found in NpC. The amplitude of the wavelike modulation of the magnetic moments also varies continuously with T below T', becoming nearly a square wave with a wavelength of six lattice spacings at $T \cong 5\,\text{K}$. The magnetic powder susceptibility χ vs. T is proportional to the magnetic moment σ measured in a constant field (Fig. 32a) and is similar to that of UAs (Fig. 14b).

In Fig. 32b one sees the drastic increase in the experimental electrical resistivity ρ as T is raised through $T' = 74\,\text{K}$. This increase appears to be much too large to be explained[40] as scattering of conduction electrons from the partial disordering of the ordered moments at T'. A most intriguing feature of the ρ vs. T data of NpP (and also NpAs) (Figs. 32b and 34) is the insignificance of the anomaly at T_N compared to that at T'. In practically all metallic antiferromagnets, the loss of magnetic ordering at the Néel temperature causes a noticeable change in slope of the ρ vs. T curve due to spin-disorder scattering. The increase of ρ at T' in NpP is about an order of magnitude larger than the increase observed for other antiferromagnets. Thus, the phase change at T' in NpP (and possibly also in NpAs) may be associated with a partial transition of conduction electrons into nonconducting or mostly localized states as T is increased (decreased in NpAs) through T'. This would give rise to the large increase (decrease in NpAs) in resistivity.

Below T', a tetragonal distortion of NpP with a $(c-a)/a$ ratio of 0.9958 ± 0.0002 (at 4.2 K) is observed, and the volume of the unit cell expands with decreasing T.[78] The latter behavior is similar to the corresponding data on UP (see Fig. 12, p. 20).

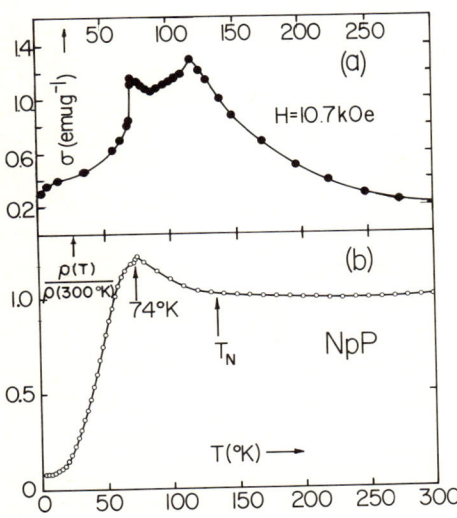

Fig. 32. Neptunium phosphide, NpP: a). Magnetization σ vs. temperature T, measured by force technique in an applied field of $H = 10.7$ kOe. b). Resistivity ρ in units of the resistivity at 300 K, as a function of T. From Ref. 90.

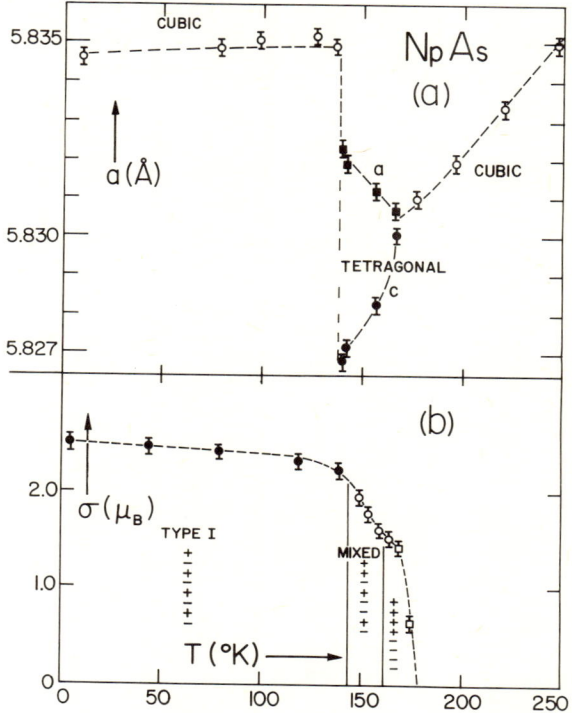

Fig. 33. Neptunium arsenide, NpAs: a). Lattice parameters as functions of temperature T. In the tetragonal phase both lattice parameters a and c are shown. Ref. 80. b). The ordered moment σ per Np atom vs. T. The orientation of the average moment of each atomic layer in one repetitive unit is indicated by the succession of + and − signs for the three different phases. From Ref. 90.

Neptunium arsenide, **NpAs**,[78,90] shows even more complex behavior. At $T_N = 175$ K there is a transition to a tetragonally distorted antiferromagnetic (4+, 4−) ordered state. At $T' = 140$ K, there is a first-order moment-jump transition to AFM-I type ordering (Figs. 33a and b), accompanied by a volume increase of $\cong 0.2\%$, a return to an undistorted cubic lattice, and a large change in the electrical resistivity (Figs. 34 and 35). The great discrepancy in the magnitude of the resistivity below 140 K as measured by the authors of References 90 and 274 is at present unexplained.

The compound **NpSb**[69] is also antiferromagnetic (see Table 3), and its magnetic susceptibility (Fig. 36) reveals a strikingly sharp peak at $T = T_N$.

2.2. NaCl-TYPE METALLIC ACTINIDE COMPOUNDS

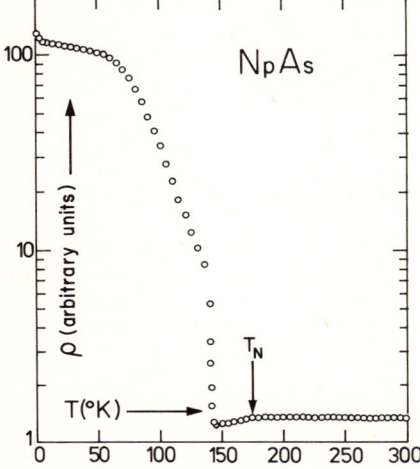

Fig. 34. Resistivity ρ of NpAs as a function of the temperature T. Note the absence of a marked contribution of spin-disorder scattering to ρ at T_N, as in Fig. 32b. From Ref. 90.

NpN is ferromagnetic with a Curie temperature $T_C \cong 98 \,\text{K}$.[91] The magnetic susceptibility[43] $\chi^{-1}(T)$ does not follow a Curie–Weiss law and the electrical resistivity[43] rises very steeply with increasing T near $T = T_N$. The ρ vs. T curve closely parallels that of MnAs near its ordering temperature, at which the transition in MnAs is thought to be first order.[15]

Neutron diffraction results[416] on single crystals show that in an applied

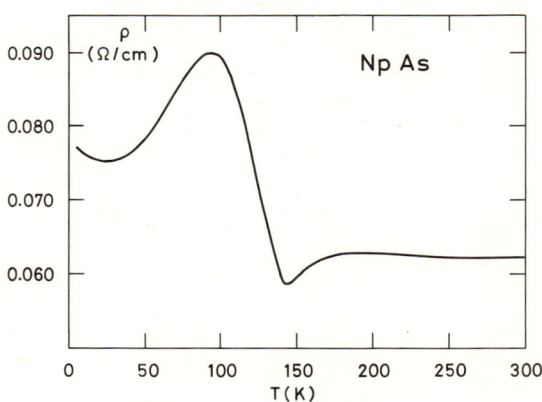

Fig. 35. Electrical resistivity of NpAs as a function of temperature T. Note the drastic (and unexplained) differences between these data and those of Fig. 34. From Ref. 274.

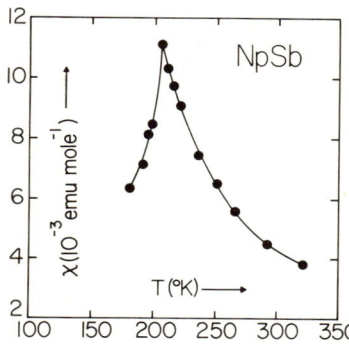

Fig. 36. Magnetic susceptibility χ of neptunium antimonide NpSb as a function of temperature T. From Ref. 91.

magnetic field of 50 kOe NpAs becomes a simple ferromagnet with $T_C = 175$ K and an ordered moment $\mu_0 = 2.5\,\mu_B$. The effective and ordered magnetic moments were explained with a localized model assuming Np^{3+} ions and negligible crystal-field splitting of their $J = 4$ ground multiplets.

2.2.9. Plutonium Compounds

Plutonium monocarbide, **PuC**, like NpC, exists as a "defect structure" defined by the formula PuC_x.[92] Magnetic susceptibility measurements on $PuC_{0.95}$[44] (Fig. 37) indicate antiferromagnetic ordering setting in at $\cong 90$ K. In this temperature range there are also anomalies in the thermoelectric power[44] and resistivity[44] (Figs. 38a and b). However, the *increase* in ρ as T is decreased below $T \cong 90$ K is highly unusual for antiferromagnetic compounds (compare, for example, Fig. 30b). Furthermore, recent specific heat measurements[72] on $PuC_{0.96}$ revealed no anomalies between 10 K and room temperature, and the data were taken at 5 mK intervals. No peaks, kinks, or jumps have been observed[93] in the lattice parameter or thermal expansion coefficient of PuC (nor PuN and Pu(C, N)) in the range 50–300 K. Thus, the nature of the transition at 90 K in PuC_x is still obscure.

The localized magnetic moments in the ordered states of the plutonium

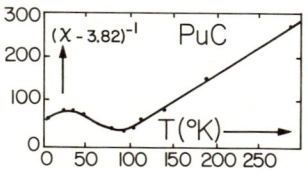

Fig. 37. Inverse magnetic susceptibility of plutonium carbide, PuC, as a function of temperature T. From Ref. 44. (Presumably in 10^{-6} emu/g.)

2.2. NaCl-TYPE METALLIC ACTINIDE COMPOUNDS

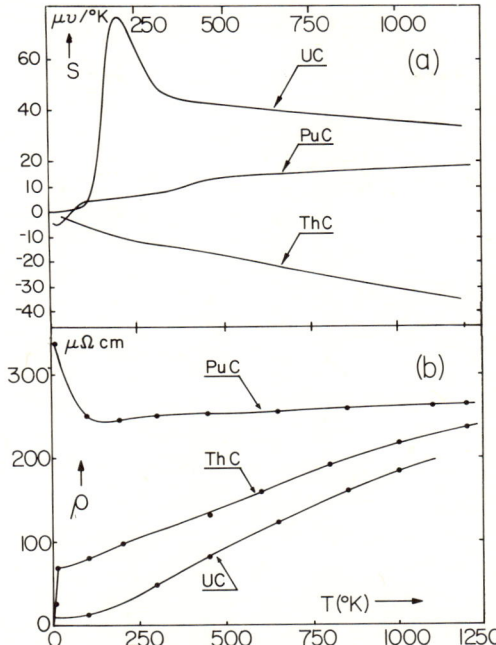

Fig. 38. Comparison of the behavior of the actinide carbides ThC, UC, and PuC. a). Thermoelectric power S vs. temperature T. b). Resistivity ρ vs. T. From Ref. 44.

monopnictides are quite small (see Table 3) and indicate strong hybridization or delocalization of the $5f$ electronic wavefunction.[92] Note that at least in the case of **PuP**, there is a negative conduction electron-spin polarization of $\cong -0.35\,\mu_B$ per Pu atom.[275] This explains the difference between the value $0.77\,\mu_B$ of the Pu moment as derived from neutron diffraction, which corresponds to the localized $5f$ moment,[275] and the value $0.42\,\mu_B$ derived from magnetization measurements,[94] which yield the *net* magnetic moment per unit cell.

The variation of the ordered moment with temperature in ferromagnetic PuP[94] is shown in Fig. 39. The magnetization curve does not follow a Brillouin function; however, the magnitude of the magnetization at $T=0\,\text{K}$ has been fitted by a localized crystal-field model assuming a $Pu^{3+}(5f^5)$ configuration.[275] The magnetic moments are aligned in the $\langle 001 \rangle$ direction, and there is a tetragonal lattice distortion[276] below T_C similar to that shown in Fig. 33 for NpAs between 140 and 175 K.

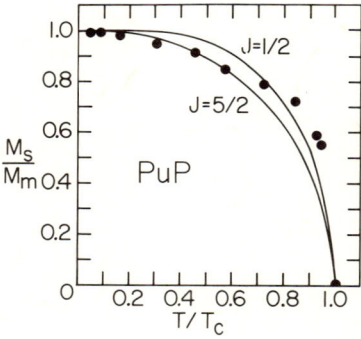

Fig. 39. Ordered moment (in units of the saturation moment) vs. temperature T (in units of T_c) for plutonium phosphide PuP. Points: experimental values. Full lines: Brillouin curves for two values of spin. From Ref. 94.

PuAs is very similar in its magnetic properties to PuP, but its σ vs. T curve can be fitted very well with a Brillouin function with $J = 5/2$,[95] which is the value of the total angular momentum of the $5f^5$ multiplet corresponding to Pu^{3+}.

X-ray photoemission studies[417] of **PuSb** show that it represents the first member of the series AcSb (where Ac = actinide atom) for which the $5f$ electrons are clearly localized. The data show the localization in two ways. First, the $5f$ part of the emission spectrum is located 1 eV below the Fermi level and shows some evidence of structure characteristic of localized atomic energy levels. Second, the inner $4f_{7/2}$ Pu level in PuSb is shifted by $\cong 5$ eV in comparison to its energy in Pu metal. The authors of Ref. 417 theorize that this shift results from the localized $5f$ electrons in PuSb being unable to screen the $4f$ electrons, whereas the itinerant $5f$ electrons in Pu metal do contribute to the screening.

PuS is a semiconductor[41] with a room temperature resistivity of $2600\,\mu\Omega$cm, which decreases exponentially to $200\,\mu\Omega$cm at $T = 900$ K. PuS apparently does not order magnetically.

The thermoelectric power and resistivity of the monocarbides of U and Th are compared to the results for PuC in Figs. 38a and b.[44,92] The anomalies in the thermoelectric power for UC are apparently not associated with a phase transition, because careful specific heat measurements failed to find any anomalies between 10 K and room temperature.[72] More detailed measurements of the thermoelectric power and of the susceptibility of pure UC would help to determine the nature of the phase transition, if there is any.

2.2.10. Solid Solutions of Uranium Monopnictides and Monochalcogenides

The compounds UAs, UP, US, and USe form solid solutions with each other, and the resulting materials (e.g., $UAs_{1-x}P_x$) retain the NaCl lattice, the

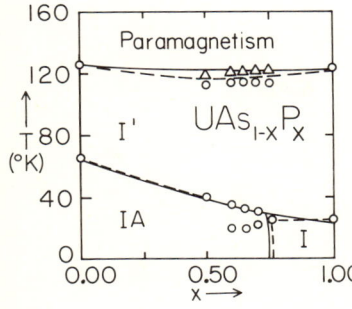

Fig. 40. Magnetic phase diagram as a function of temperature T and x of the solid compounds $UAs_{1-x}P_x$. I, IA: types of magnetic ordering (see Fig. 4). I′: low-moment phase of type I ordering. The circles and triangles are data of Ref. 3, and the solid curves are the result of the theory of Ref. 16 (see Chapter 3).

different anion species being randomly distributed over the anion sublattice. The changes in the type of magnetic ordering and in the magnitude of the ordered moment as functions of temperature T and relative composition x have been investigated by neutron diffraction. The results shown in Figs. 40–44 and 46–47 as magnetic phase diagrams in the x–T plane reveal complicated behavior.

In the case of the $UAs_{1-x}P_x$[3,60] system (Fig. 40), the interval $0.6 < x < 0.75$ represents a region of sample inhomogeneity in which the reflections characteristic of both the AFM-IA (see Fig. 4) and the AFM-I′ (i.e., the low-moment state of AFM-I) phases were observed. Clearly, more data are needed to clarify the phase boundaries.

The addition of only about 3% US to UP destroys the AFM-I magnetic phase at $T = 0$ K, as Fig. 41 shows.[51,52,96,97] The Néel temperature T_N decreases by about 25% as x increases from $x = 0$ (UP) to $x \cong 0.2$ in $UP_{1-x}S_x$,[51] in contrast to the results for $UAs_{1-x}P_x$, for which T_N is nearly independent of x.[3] This contrast can be understood from the fact that UAs and UP are magnetically very similar, whereas US is very different from both these compounds. The region marked A is an antiphase structure[10] consisting of ferromagnetic layers

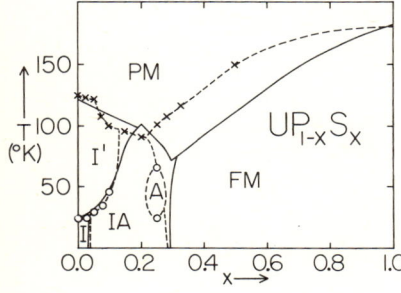

Fig. 41. Magnetic phase diagram as a function of T and x of the solid compounds $UP_{1-x}S_x$. I, IA, I′: same meaning as in Fig. 40. A: antiphase magnetic ordering discussed in text, PM = paramagnetism, FM = ferromagnetism. The circles and crosses are experimental data of Ref. 51, and the solid curve is the theoretical result of Ref. 16. The dashed curves connect data points.

stacked in the sequence 5+, 4−, 5+, 4−, ..., where the symbols denote that the moments of the corresponding number of layers point up (+) or down (−) with respect to a crystal axis perpendicular to the plane of the layers. The experimental data are not sufficient to delineate the region in the phase diagram occupied by the A phase. The magnetization *vs.* T curve of $UP_{0.75}S_{0.25}$ is shown in Fig. 42, and one sees the customary moment-jump transition at the temperature $T' = 20$ K where the AFM-IA magnetic phase transforms to the A phase.[10] Of equal interest is the transition at $T \cong 70$ K to the FM phase, which is *not* accompanied by any discernible discontinuity in the ordered moment μ.[52]

The materials $UP_{0.9}S_{0.1}$ and $UP_{0.8}S_{0.2}$ also exhibit transitions between different types of magnetic ordering without corresponding discontinuities in μ (see Fig. 43).[52]

In $UP_{0.9}S_{0.1}$, the moment-jump transition at $T' = 55$ K from the AFM-IA phase ($T < T'$) to the AFM-I phase ($T > T'$) corresponds to a jump in the susceptibility[52] χ similar to that seen in the data on UP and UAs (see Figs. 6b and 14b). At a higher temperature $T'' \cong 90$ K, the AFM-I phase transforms back to the AFM-IA phase with no discontinuity in the ordered moment. A surprising observation is that the latter transition is associated with a sharp susceptibility peak of a width of $\cong 10$ K. A similar but broader peak in the susceptibility[52] of $UP_{0.8}S_{0.2}$ at a temperature of $\cong 90$ K has been observed. These peaks have been attributed to ferromagnetic transitions induced by the applied magnetic field of 12 kG used in making the measurements.[52]

Between $x = 0.2$ and $x = 0.35$, there are intervals of temperature where neutron reflections characteristic of several types of magnetic structures are simultaneously observed[51,52,96,97] in $UP_{1-x}S_x$. This behavior may result from sample inhomogeneity.

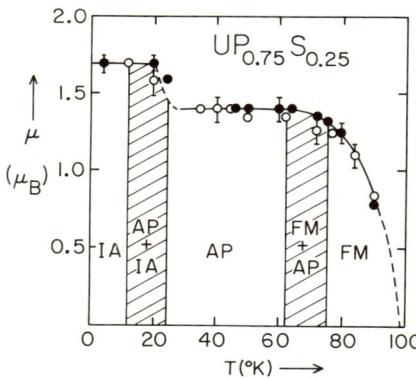

Fig. 42. The ordered magnetic moment μ per U ion *vs.* T of $UP_{0.75}S_{0.25}$ according to the experiments of Ref. 10. The shaded regions are temperature intervals in which two magnetic phases coexist. The symbols IA, AP, FM, and PM denote different magnetic phases (see Fig. 41).

2.2. NaCl-TYPE METALLIC ACTINIDE COMPOUNDS

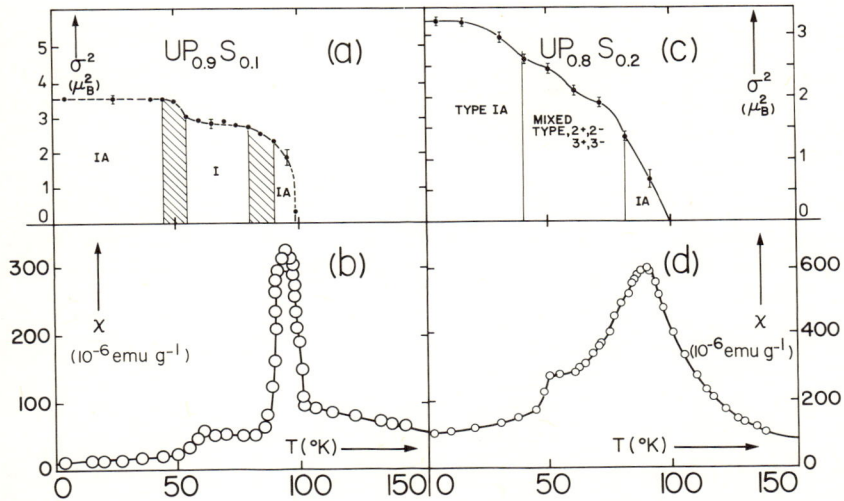

Fig. 43. The square of the ordered moment σ per U ion [(a) and (c)] and the gram-magnetic susceptibility χ [(b) and (d)] vs. T of $UP_{0.9}S_{0.1}$ [(a) and (b)] and $UP_{0.8}S_{0.2}$ [(c) and (d)] from Ref. 52. The shading indicates the coexistence of two magnetic phases. The notation 2^+, 2^- indicates a repetitive structure formed by two upspin layers followed by two downspin layers.

In the $UAs_{1-x}S_x$ phase diagram[2,98] (Fig. 44), the region marked L represents a longitudinal wave arrangement of the magnetic moments similar to that found in NpP discussed earlier. The magnetic structures and transitions observed in $UAs_{1-x}S_x$ are similar to those already discussed in the case of $UP_{1-x}S_x$.

Measurements of the electrical resistivity ρ of $U_{1-x}Th_xS$[99] and the electronic specific heat γ of UP_xS_{1-x}[100] show sharp maxima as x is varied (see Fig. 45). These maxima were interpreted[45] as resulting from the proximity in energy of narrow virtually bound f-levels to the Fermi level. However, in $UP_{1-x}S_x$ the maximum in γ was not found to be associated with a maximum in the resistivity as a function of x.[55]

The $UAs_{1-x}Se_x$ phase diagram[60,101] (Fig. 46) seems very similar to that of $UAs_{1-x}S_x$, but clearly more data are needed for its completion.

The $UP_{1-x}Se_x$ phase diagram[260] (Fig. 47) reveals the interesting new feature of moment-jump transitions at $T' \cong 58\,K$ within the AFM-IA phase existing in the composition range $x \cong 0.1$–0.2. This contrasts with UAs as well as with the other phase diagrams, where the moment jump in the IA phase is always associated with a change in the type of magnetic ordering.

The effect of the replacement of U by the nonmagnetic Th has been

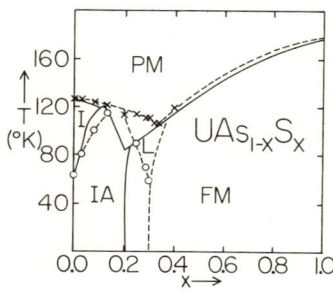

Fig. 44. Magnetic phase diagram of $UAs_{1-x}S_x$. L: longitudinal wave magnetic ordering explained in text. The circles and crosses are data of Ref. 2, and the other curves and symbols have the same meanings as in Figs. 40 and 41.

studied through magnetization measurements.[102] In $U_xTh_{1-x}Sb$, at $x = 0.8$ a ferromagnetic order is observed with strong anisotropy favoring ⟨111⟩-alignment, and with a saturation moment μ_s which is the same as in pure USb. Saturation is reached at 60 kOe at 4.2 K. Between 60 and 140 kOe, the moment in the ⟨100⟩ direction stays constant at $\frac{1}{3}\mu_s$, which shows that a very high anisotropy field is holding the moments in the ⟨111⟩ direction. The change from AFM-I to FM behavior through the dilution with Th is probably due to the change in the RKKY interaction through the addition of one conduction electron per U atom replaced by a Th atom (see Section 3.1.3e).

At the composition $x = 0.1$, the compound is paramagnetic down to 4.2 K. The magnetization induced in fields of up to 100 kOe (0.6 μ_B/U atom) is

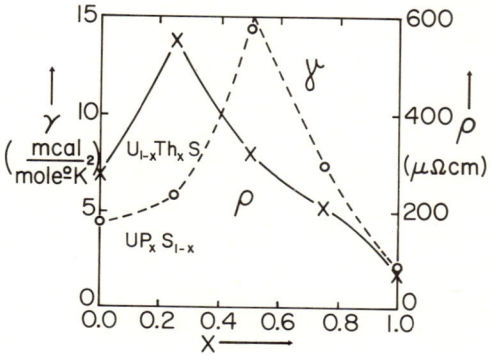

Fig. 45. The electronic specific heat coefficient γ vs. x in UP_xS_{1-x} (circles and dashed curves, Ref. 100) and the electrical resistivity ρ vs. x in $U_{1-x}Th_xS$ (crosses and solid curves, Ref. 99). The sharp peaks are thought to result from narrow f-levels moving through the Fermi energy level as x is varied.

2.2. NaCl-TYPE METALLIC ACTINIDE COMPOUNDS 49

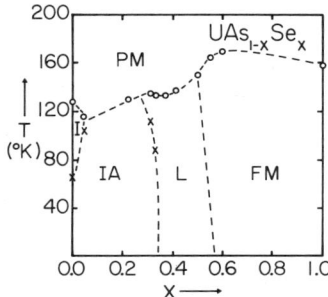

Fig. 46. Magnetic phase diagram of $UAs_{1-x}Se_x$. The circles and crosses are data of Ref. 101, and the other symbols have the same meaning as in Fig. 41.

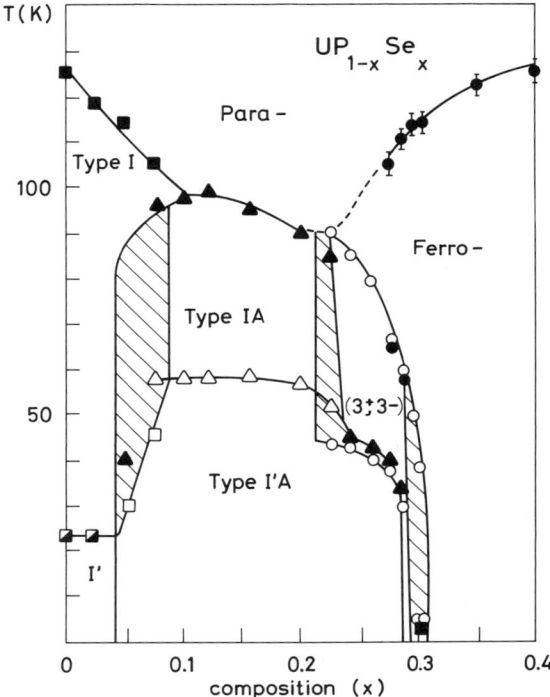

Fig. 47. Magnetic phase diagram for the $UP_{1-x}Se_x$ system established on the basis of susceptibility and neutron diffraction data. The shaded areas denote two-phase regions. From Ref. 260.

smaller than the authors[102] would expect for any $5f^3$ (U^{3+}) crystal-field-level scheme, which they assume for the $x = 0.8$ composition. They conclude that the only explanation possible within a crystal-field model is to assume a Γ_1 single ground state within a U^{4+} ($5f^2$) configuration.

Recent neutron diffraction studies[418] of $U_x Th_{1-x} Sb$ show that the magnetic ordered moment increases linearly between $x = 0.2$ and $x = 0.8$, indicating that a change of valence from U^{4+} to U^{3+} occurs over the same interval of x. If this interpretation is correct, this compound would be an interesting example of a mixed-valence actinide material.

The same authors[418] find that the magnetic moment per U ion in the isostructural compound $U_x Y_{1-x} Sb$ increases linearly between $x = 0.2$ and $x = 0.5$, at which point the moment jumps from $2.15 \pm 0.25 \mu_B$ to the value $2.8 \mu_B$ and remains constant from $x = 0.5$ to $x = 1.0$. The compound is paramagnetic from $x = 0.0$ to $x \cong 0.2$, ferromagnetic from $x \cong 0.2$ to $x \cong 0.45$, and has the AFM-I triple-k structure for $x \gtrsim 0.45$.

Magnetization and neutron diffraction studies have been done on single crystals of $U Sb_{0.9} Te_{0.1}$[103] and $U Sb_{1-x} Te_x$.[267] At 1.5 K a saturation moment of $0.92 \mu_B$ develops in the $\langle 111 \rangle$ direction in an applied field >40 kOe. A moment of $\cong 0.85 \mu_B$ remains after removal of the field. In the neutron diffraction experiment, in addition to the ferromagnetic ($\mathbf{k} = 0$) component of amplitude $A_0 = (0.93 \pm 0.1) \mu_B$, reflections corresponding to three wave vectors $\mathbf{k}_1 = (2/3, 0, 0)$, $\mathbf{k}_2 = (0, 2/3, 0)$, and $\mathbf{k}_3 = (0, 0, 2/3)$ are also observed. These yield three Fourier components $\mathbf{m}(\mathbf{k}_i)$ ($i = 1, 2, 3$) of the magnetization with amplitudes $A(\mathbf{k}_i) = (2.33 \pm 0.08) \mu_B$, and directions parallel to $\mathbf{k}_i$.

The authors[103] interpret these results in terms of a structure in which all U moments of $2.9 \mu_B$ lie along the $\langle 111 \rangle$ axis, but are simultaneously modulated along the three cube axes in the sequence $++-$. This gives $\mu_s = \frac{1}{3} \mu_{ion}$ as observed. Above 160 K the ferromagnetic component of the moment disappears, and an incommensurate multiaxial structure develops, which cannot be interpreted unambiguously. At $T_N = 205$ K the crystals become paramagnetic.

As in $U_x Th_{1-x} Sb$ (see above, this subsection), it is observed that a smaller field ought to be applied in a *hard* direction rather than in the easy $\langle 111 \rangle$ direction to achieve saturation (which occurs when the crystal becomes a single domain). This phenomenon is as yet unexplained, since the rotation of domain magnetization is unlikely to be due to the extremely high (probably >160 kOe) anisotropy fields.

The authors[103] point out that the direction of the uranium moments, parallel to $\langle 111 \rangle$ in $USb_{0.9} Te_{0.1}$, is in apparent conflict with the Type I-AFM structure of USb, where the moments are supposed to be in the $\langle 100 \rangle$ direction,

2.3. UX₂ AND UXY TETRAGONAL URANIUM COMPOUNDS

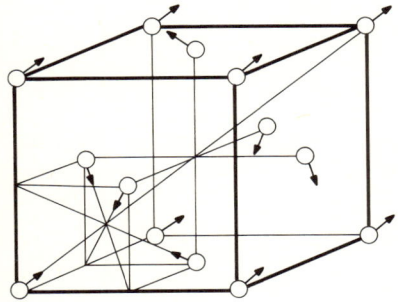

Fig. 48. Model of a triaxial antiferromagnetic structure in the face-centered cubic lattice. The dots represent the magnetic ions, whose moments (arrows) point in the $(\pm 1, \pm 1, \pm 1)$ directions. Originally proposed for $(Ni,Fe)_3Mn$ alloys,[104] this structure is envisaged for uranium monopnictides. From Ref. 103.

alternating in sign in successive planes perpendicular to that direction. They suggest that the conventional AFM-I type interpretation of neutron scattering experiments on U-pnictides may generally be wrong and emphasize that the multiaxial structure proposed originally for Mn alloys[104] is also compatible with the neutron scattering data. In the latter structure the moments alternate between the eight possible $\langle \pm 1, \pm 1, \pm 1 \rangle$ directions (see Fig. 48).

It is as yet premature to judge this interpretation, but we remark that USb and $USb_{0.9}Te_{0.1}$ may have completely different structures, just as USb and $U_{0.8}Th_{0.2}Sb$ have, one being AFM-I, the other a FM with the full U moment. Also, the monopnictide UN exhibits a tetragonal distortion[68] in the ordered phase which points to the AFM-I structure rather than to the triaxial structure, in which no cubic axis is a preferred direction for strain.

If our knowledge of the magnetic structure of U-monopnictides requires revision, the present interpretation of the magnitude of U moments as well as of the magnetic phase transitions has to be reexamined. Hence, it is important that the magnetic structure of the monopnictides be clarified.

2.3. UX₂-Type and UXY-Type Tetragonal Uranium Compounds

In these compounds X and Y are elements of groups VA or VI of the periodic table. Their physical properties have been elucidated mostly by the Polish researchers at Wroclaw,[21] and the known facts are summarized in Table 4. The crystal structure is tetragonal, most of the compounds having the space group $D_{4h}^7(P4/nmm)$. The UX₂ and UXY compounds are isomorphic to Cu_2Sb and PbFCl, respectively, with the exceptions of US₂, USe₂, and UAsTe (see Table 4). Like the NaCl-type metallic actinides discussed in the previous section,

Table 4. Tetragonal Conductor or Semiconductor Compounds of the Type AX_2 and AXY (A = Actinide). (a, c: lattice parameters. Other symbols are defined in Table 3, p. 14. The meaning of the asterisk is explained on p. 55.)

Compound AX_2, AXY	a (Å)	c (Å)	Symmetry	Order	T_N or T_C (K)	μ_0 (μ_B)	μ_p (μ_B)	θ (K)	Refs.
UP$_2$	3.810	7.764	P4/nmm	+−−+	203	1.0	2.0; 2.5*	94; 86*	106, 110, 114
UAs$_2$	3.962	8.134	P4/nmm	+−−+	272	1.6	2.4; 2.94	72; 34	107, 109, 117, 116
USb$_2$	4.283	8.758	P4/nmm	+−−+	203	0.94	2.47; 3.04	72; 18	107, 109, 117, 116
UBi$_2$	4.445	8.908	P4/nmm	+−+−	183	2.1	2.85; 3.4	0; −53	109, 117
α-US$_2$	10.312	6.352	I4/mcm		~20		3.6*	−120*	74, 121
α-USe$_2$	10.697	6.606	I4/mcm		~13		3.2*	−48*	74, 121
UOSe	3.904	6.979	P4/nmm	++−−	≤75	2.21	2.53; 3.3*	0; −120*	105, 117
UOTe			P4/nmm	+−+−	164	2.1	2.57	0; −56	106, 117
UOS	2.835	6.685	P4/nmm	++−−	55	2.0	2.2*; 2.57	0; −100*	110, 117
UAsS	3.884	8.176	P4/nmm	++++	124	1.14	3.34	−380	119, 120
UAsSe	3.962	8.422	P4/nmm	++++	113	1.36	3.41	−195	119, 120
UAsTe	4.1483	17.2538	I4/mmm	++++	66	1.29	3.34	−95	119, 120
UNSe			P4/nmm	++++	88		2.03	43	117
UNTe			P4/nmm	++++	59		1.87	44	117

2.3. UX_2 AND UXY TETRAGONAL URANIUM COMPOUNDS

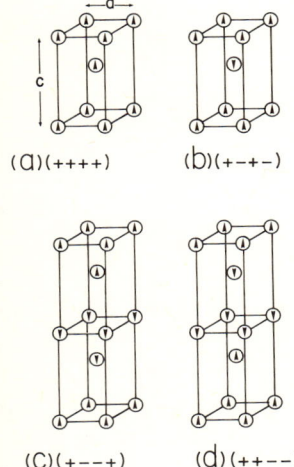

Fig. 49. Four types of magnetic ordering found experimentally in tetragonal compounds of the classes UX_2 and UOX. The open circles represent the U ions and the arrows show the directions of the ordered moments. Each type of ordering is labelled by a sequence of plus and minus signs as shown, indicating the sign of the moment with respect to the vertical axis in consecutive horizontal layers.

the tetragonal compounds are mostly good conductors of heat and electricity, having room temperature resistivities of $\cong 200\text{-}500\,\mu\Omega\text{cm}$.[105-107]

Neutron diffraction experiments[105,108-112] show magnetic ordering with the magnitudes μ_0 of the ordered moments in the range $1.0\text{-}2.0\,\mu_B$. The four types of magnetic ordering found are illustrated and defined in Fig. 49. Przystawa and Suski[113] have calculated which of these four phases is stable when the exchange interactions between several near neighbor actinide ions are specified. There are indications[109,114,115] that long-range magnetic exchange interactions via conduction electrons may be important for the tetragonal compounds, and that as a result the calculated stability criteria may not be valid. Unfortunately, the neutron diffraction experiments have been performed only at two or three temperatures, so that no complete curves of sublattice magnetization vs. T are available. Thus, we do not know whether all these materials have first-order magnetic phase transitions.

Precise specific heat measurements[116] (Fig. 50) have been done on UAs_2 and USb_2 samples. They reveal very pointed Lambda-type peaks (similar to those shown in Fig. 39 for U_3P_4) at 272.5 K and at 202.5 K, respectively, which indicate a first-order transition between the ordered $(+--+)$ and the paramagnetic states. The magnetic contribution to the entropy is estimated to be $\Delta S(\text{mag}) = 0.99\,\text{cal/K mole}$ in UAs_2 and $\Delta S(\text{mag}) = 170\,\text{cal/K mole}$ in USb_2. That these values are significantly lower than the spin-only magnetic entropy $R \ln 3 = 2.18\,\text{cal/K mole}$ is a rather general feature of many actinide compounds. The electronic specific heat coefficients are $\gamma(UAs_2) = 1.36 \times 10^{-3}$ and $\gamma(USb_2) = 2.98 \times 10^{-3}\,\text{cal K}^{-2}\text{mole}^{-1}$.

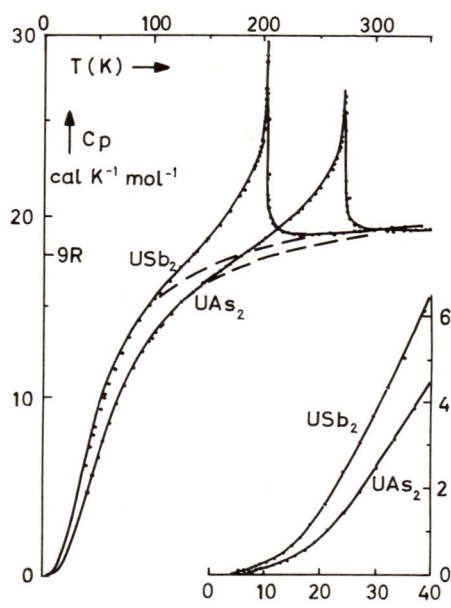

Fig. 50. Heat capacity c_p vs. temperature T of USb$_2$ and UAs$_2$ powder samples: points and full lines. Dashed lines: calculated lattice heat capacities. The peaks are at 272.2 and 202.5 K for UAs$_2$ and USb$_2$, respectively. From Ref. 115.

Besides the peaks corresponding to the magnetic ordering listed in Table 4 for the dipnictides UX$_2$, additional peaks have been found[117] in the neighborhood of 10 K. It is not known whether these correspond to new phase transitions.

Some data are available on the magnetic susceptibility of **UOS**.[111] The susceptibility vs. temperature of the mixed uranium dipnictides UXX' (X, X' = N, P, As) and of **USnTe** have been measured[117] in the range of 4.2–900 K. Transitions to the antiferromagnetic state have been observed in all of these compounds.[117]

The compounds **UGeY** (Y = S, Se, Te) order antiferromagnetically. The magnetic moments of the U ions at 4 K are 1.26, 0.26, and 1.50 μ_B, and the Néel temperatures 88, 40, and 73 K, for the three compounds, respectively.[118] The magnetic moments are found by neutron diffraction[118] to be along the c axis, and ferromagnetic sheets perpendicular to the c axis have the sequence (++−−) in the sulphide and telluride, (+−+−) in the selenide. The authors do not attempt to theoretically deduce the observed magnetic moments from crystal-field theory because of the complication due to low symmetry.

The magnetization of ferromagnetic **UAsS**, **UAsSe**, and **UAsTe** has been measured in the temperature range between 3 K and their respective Curie points in fields up to 80 kOe.[119,120] They exhibit large anisotropy fields on the order of 10^7 erg/cm^3. The magnetic susceptibility in the paramagnetic state[119] has a temperature dependence reminiscent of that of ferromagnets. However, the presence of uranium atoms of different valency in metallic compounds has not been proved so far.

Unusual magnetic and electronic properties have been found, as Figs. 51–53 and 55 indicate, in some other ternary and binary compounds. First, the inverse paramagnetic susceptibility $\chi^{-1}(T)$ of several of these compounds exhibits a sharp change in slope at $T^0 \cong 300$ K.[105–108] The Curie–Weiss law is obeyed fairly well above and below T^0, but with two different values of the effective paramagnetic moment μ_p below and above T^0 differing by about 10%. In the case of **UOSe** (Fig. 51), the change is from $\mu_p = 2.87 \mu_B$ $(T < T^0)$ to $\mu_p = 3.3 \mu_B$ $(T > T^0)$.[105] The cause of the slope change is unexplained; it seems too sharp to be due to thermal population of higher crystal-field energy levels. There often appear discrepancies in the reported values of μ_p and the paramagnetic Curie temperature θ, because some experimenters use the lower and some the higher temperature portions of the $\chi^{-1}(T)$ curves in determining these quantities. In Table 4, the data for $T > T^0$ are used to find μ_p and θ, and an asterisk is added to indicate that μ_p and θ change at T^0.

The resistivity data[105] on UOSe (Fig. 52) show that it is a semiconductor, and a large change in the semiconducting energy gap at the Néel temperature $T_N \cong 90$ K can be deduced. The broad maximum in the $\log \rho$ vs. T data at T_N is also interesting. A similar maximum in ρ for current flow parallel to the c axis is seen in the case of **UP$_2$**[106] (Fig. 53) at about 110 K, and this is well below the Néel temperature $T_N = 203$ K of UP$_2$. Henkie and Trzebiatowski[106] have suggested a temperature-dependent displacement of the U ions in the c direction giving rise to a new gap and thereby to a resistance anomaly in the c direction.

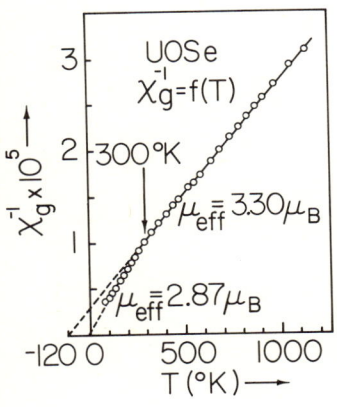

Fig. 51. The inverse magnetic gram-susceptibility χ_g^{-1} vs. T of UOSe, from Ref. 105. The value of the effective paramagnetic moment μ_{eff} deduced from the slope of χ_g^{-1} changes abruptly at $T = 300$ K, as shown.

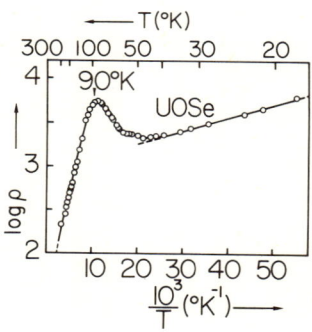

Fig. 52. The logarithm of the electrical resistivity ρ vs. T (upper scale) and vs. $10^3/T$ (lower scale) for UOSe, Ref. 105. The anomaly at 90 K is associated with antiferromagnetic ordering.

The compounds US_2,[74,121] USe_2,[122] and UTe_2[123] are of interest as possible singlet ground state induced-moment systems.[95] There are two crystal phases, the tetragonal α-phase listed in Table 4 and the orthorhombic β-phase.[125] Assuming the U^{4+} valence state of the uranium ions, Suski et al.[125] have calculated that the $^3H_4$ term (see Chapter 3 for an introduction to crystal-field theory) splits up in the low-symmetry crystal field of these compounds into nine singlets, the two singlets lowest in energy being very closely spaced (Fig. 54). Even though the magnetic moment of the singlet ground state is zero, Wang and Cooper[124] have shown that a magnetically ordered state can arise in such a

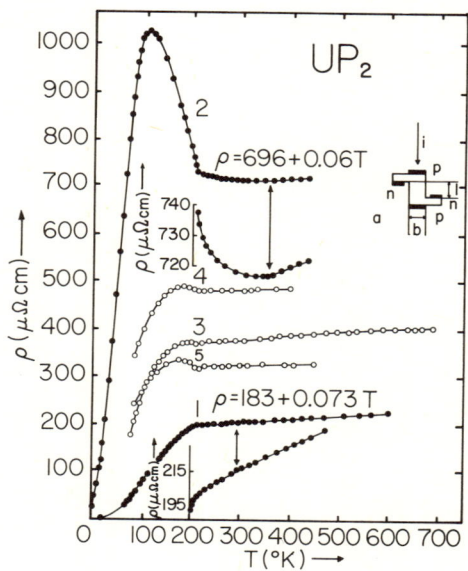

Fig. 53. The electrical resistivity ρ vs. T of several samples of UP_2, after Ref. 106. The inserted scales show details of behavior of samples 1 and 2. The inserted sketch (a) shows the experimental arrangement. The equations represent best linear fit to portions of the curves. 1: single crystal $\perp c$ axis; 2: single crystal $\parallel c$ axis; 3, 4, 5: different polycrystalline samples.

2.3. UX$_2$ AND UXY TETRAGONAL URANIUM COMPOUNDS

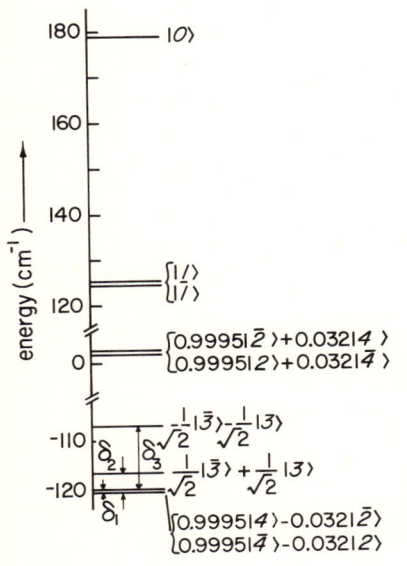

Fig. 54. The crystal-field energy levels and wave functions of the U^{4+} ion in the orthorhombic β-phase of UX$_2$ compounds, as calculated in Ref. 125. Shown are the nine singlets arising from the $^3H_4$ multiplet. The singlets are linear combinations of the states $|M_J\rangle$, $M_J = -J, -J+1, \ldots, J$. The bar over a number indicates its negative. The δ_i ($i = 1, 2, 3$) are splittings.

system at low temperature if the off-diagonal Zeeman matrix element between the two lowest singlets is sufficiently large. The magnetic powder susceptibility[125] of β-US$_2$ at low temperatures (Fig. 55) shows behavior similar to that predicted by the Wang–Cooper theory.

The ternary chalcogenide compounds **UNSe** and **UNTe** are also found to be ferromagnetic,[117] with Curie points of 88 K and 59 K, respectively. The dependence of the reciprocal magnetic susceptibility on temperature shows a convex curvature,[117] which is being ascribed[117] to the effect of the crystal field on the U^{4+} ion, thought to be in a doublet ground state in the C_{4v} symmetry.

NpAs$_2$, which is isostructural to UAs$_2$, has been reported[126] to be magnetically ordered, whereas **PaAs$_2$** exhibits[127] a temperature-independent paramagnetism between 35 and 300 K. Their crystal structure is $P4/nmm$. For PaAs$_2$, a = 3.978 Å, c = 8.154 Å. Neutron diffraction measurements[434,435] on a single crystal of NpAs$_2$ reveal an unusual type of antiferromagnetic ordering below T_N = 54 K. The Np magnetic moments are parallel to the c axis but their magnitudes are modulated sinusoidally in the [100] direction. At T = 18 K there is a first-order transition to ferromagnetism, the moments having a magnitude of 1.45 μ_B and being aligned along the c axis.

Neutron diffraction, NMR, or Mössbauer studies to determine the temperature dependence of the sublattice magnetization of the UX$_2$ and UXY

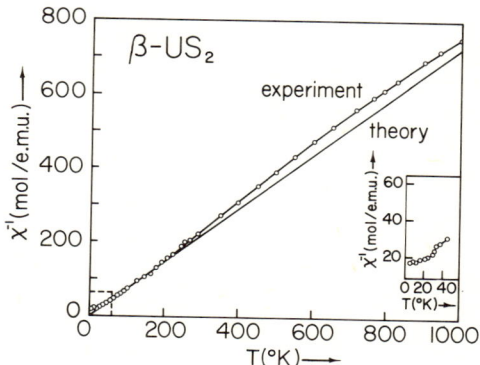

Fig. 55. The inverse molar susceptibility χ^{-1} vs. T of β-US$_2$ Ref. 125. The insert shows details near $T = 20$ K. The theory (lower curve) is based on the crystal-field model shown in Fig. 54.

compounds would be useful. The specific heat of these materials should also be determined. Theoretical calculations of the wavefunctions of the actinide ions have not yet been carried out in detail for the crystalline environments of the tetragonal uranium compounds. The knowledge of these wavefunctions would be helpful to compare crystal-field theory predictions with experiment. A preliminary calculation[419] based on the point-charge crystal-field model of NpX$_2$ compounds failed to explain the observed values of the ordered magnetic moments.

2.4. A$_3$X$_4$-Type Metallic Actinide Compounds

The compounds having the formula U$_3$X$_4$ (X = P, As, Sb, Bi, Se, and Te) are ferromagnetic and have the body-centered cubic Th$_3$P$_4$ structure (see Table 5). The magnetic moments are thought to align themselves along the [100], [010], and [001] axes, with the easy axis and the resultant magnetization pointed in the [111] direction.[29,128] Thus, the magnetic structure appears to have three sublattices, as had been theoretically predicted,[129–131] but a neutron diffraction study to resolve the structure is clearly needed.

The anisotropy of the magnetization in the (110) plane of U$_3$P$_4$ is shown in Fig. 56 and similar results are obtained for U$_3$As$_4$.[29] The magnetization of U$_3$P$_4$ was observed[421] to increase steadily from 18 emu/g to 26 emu/g as a magnetic field in the [100] direction (hard axis) was increased from 0 to 500 kOe at $T = 4.2$ K. The latter magnetization is 90% of the saturation value measured in the [111] direction. The temperature dependence of the magnetization of U_3P_4 and U_3As_4 single crystals could not be explained by simple spin-wave theory.[422] The specific heat c_p as a function of T of U_3P_4 is shown in Fig. 57, and the entropy ΔS of magnetic disordering was found to be $\Delta S = 1.03$ cal/deg-mole.[132] This result is close to the value $\Delta S = R(\ln 2) = 1.38$ cal/deg-mole expected for a magnetic doublet and lends some support to the crystal-field model of Troć et al.,[133] which assumes a ground state for U^{4+} consisting of two closely spaced singlets (i.e., a pseudodoublet). Recently, the crystal-field model has been extended by the addition of p-f mixing; i.e., the quantum-mechanical overlap between f and p orbitals.[423] This enables the authors to explain the nature of the spin structure and the orientational transition in a high field. However, the anomalously small ($\cong 1\%$) increase in the U magnetic moment under high applied fields ($\gtrsim 200$ kOe) is unexplained.[423] The inverse paramagnetic susceptibilities of the U_3X_4 compounds display noticeable curvature as functions of T.[133]

The electrical properties of these compounds are very interesting. All are semimetallic with resistivities of several hundred $\mu\Omega$cm that are highly anisotropic in the ordered state.[134,135] Especially unusual is the magnetic resistivity ρ_m of U_3As_4 (Fig. 58), which shows only a weak anomaly at $T_C = 198$ K and remains relatively constant with decreasing T down to $T \cong 110$ K before falling off rapidly.[136] This contrasts with the normal behavior of magnetically ordered metals, wherein ρ_m decreases sharply just below the ordering temperature (see e.g., Fig. 23). The anomaly has been investigated on a finer scale,[137] and it has been found that besides a tiny peak at T_N, a broad, but also small, maximum develops in the longitudinal resistivity of a single crystal of U_3As_4 in very small fields (5 → 60 Oe), and that this maximum shifts to lower temperatures (200 → 175 K) with increasing field. The Hall coefficient is also shown on the

Table 5. A_3X_4-Type Compounds of Body-Centered Cubic Structure. They are Metallic Ferromagnets. (Symbols are defined in Table 3, p. 14.)

Compound U_3X_4	a_0 (Å)	T_C (K)	θ (K)	μ_0 (μ_B)	μ_p (μ_B)	Refs.
U_3P_4	8.214	138	140	1.55	2.75	128, 133, 29
U_3As_4	8.521	198	200	1.71	2.94	133, 29, 21
U_3Sb_4	9.112	146	155	1.6	3.04	133, 21
U_3Bi_4	9.368	108	110	1.5	3.14	133, 21
U_3Se_4	8.760	130	−80	0.35	3.1	133, 139
U_3Te_4	9.412	120	40	0.44	3.14	133, 139

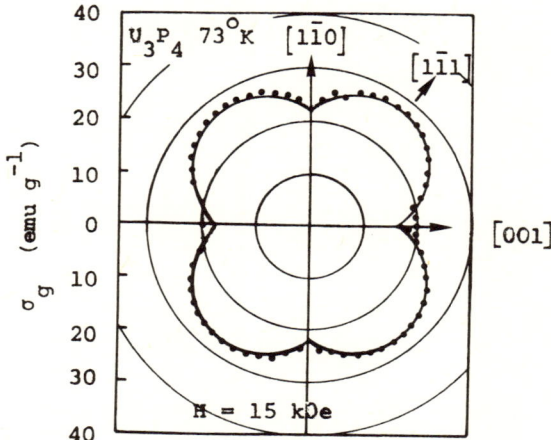

Fig. 56. The gram-magnetization σ_g of U_3P_4 measured at $T = 73$ K vs. the direction of the field $H = 15$ kOe, applied in the basal plane, Ref. 29. The data show the effect of the cubic crystalline anisotropy.

same diagram (Fig. 58). The thermoelectric power of U_3As_4 also varies anomalously below T_C (Fig. 59).[136] The thermoelectric power of U_3P_4 (Fig. 59) is of the order of magnitude usually observed in metals and shows a definite kink at the Curie temperature.

The value 0.5 for the number of carrier electrons per U atom in U_3P_4 has

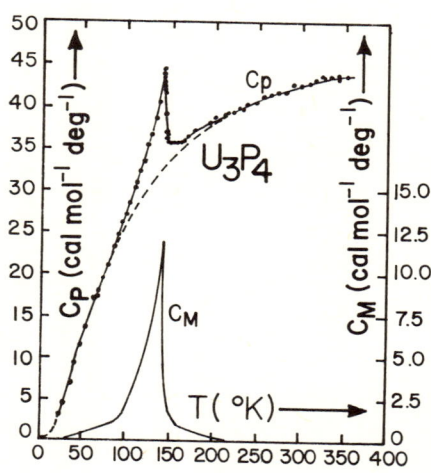

Fig. 57. The molar heat capacity c_p (upper data) and the derived magnetic heat capacity c_M (lower curve) vs. T of U_3P_4, Ref. 132.

2.4. A_3X_4-TYPE METALLIC ACTINIDE COMPOUNDS

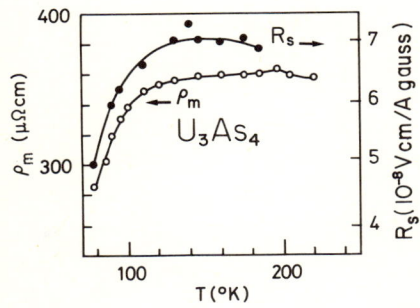

Fig. 58. The magnetic part of the electrical resistivity ρ_m and the Hall coefficient R_s vs. T in U_3As_4, Ref. 136. The small size of the anomaly in ρ_m near $T_c = 198$ K is unusual.

been deduced from optical reflectivity measurements[438] and is consistent with the Hall effect data.[136] The electrical resistivity, thermopower, and Hall effect data recently obtained[437] for U_3Sb_4 resemble closely the corresponding data on U_3As_4.

There is a distortion of about 0.18% along the [111] direction in U_3P_4 setting in below T_C,[138] as was found for US.[72] A temperature-dependent distortion is thought to be responsible for the unusual magnetization vs. T curves found in U_3Te_4 and U_3Se_4.[74, 139, 140]

The compound Pu_3S_4 is antiferromagnetic[141] ($T_N = 10$ K). In Am_3Se_4 and $AmTe_4$ no ordering has been found above 4.2 K.[142]

Np_3As_4 has been found[416] by magnetization measurements to be ferromagnetic with $T_C = 81$ K, a saturation moment $\mu_s = 1.33 \mu_B$ per Np ion, an effective moment $\mu_p = 1.64 \mu_B$, and a paramagnetic Curie constant $\theta = 83$ K. The determination of the temperature dependence of the ordered magnetic moment by Mössbauer spectroscopy yielded[420] a normal Brillouin-type curve, with an ordered moment of $1.7 \mu_B$ at $T = 0$. The isomer shift indicated that the Np ions are in the $Np^{4+}(5f^3)$ configuration.

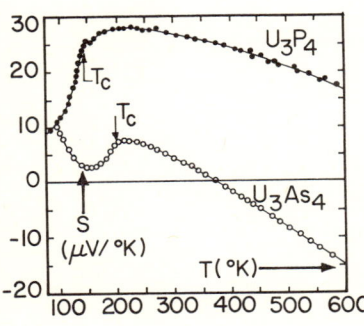

Fig. 59. The thermoelectric power S vs. T of U_3P_4 and U_3As_4, Ref. 136. T_c = Curie temperature. These data show a closer relationship to the magnetic ordering than the data of Fig. 58.

2.5. Intermetallic Actinide Compounds

2.5.1. General

The compounds of actinide elements with other metallic elements have been intensively studied because of the unusual temperature dependencies of their electrical resistivities and of their magnetic susceptibilities (see Figs. 60–62). These materials fall into two main classes — those having the formula AX_2 (A = actinide; X = metal or semiconducting element) with the cubic $MgCu_2$ crystal structure, and those of the form AX_3 with the face-centered cubic $AuCu_3$ lattice.[143] Tables 6 and 7 summarize the physical properties of a representative number of these compounds.

In a number of examples, such as UAl_2,[18,144,145] $PuAl_2$,[18] $PuPt_2$[146,147] (Fig. 60), and several UX_3 compounds[147] (Fig. 61), the electrical resistivity ρ falls off with decreasing temperature. In some of these compounds, e.g., in $PuAl_2$, a critical temperature T_s may be observed below which the decrease of ρ is very steep. The decrease of ρ somewhat resembles that resulting from the onset of magnetic ordering in other metallic compounds, but strangely no other evidence of magnetic ordering (e.g., from neutron diffraction or susceptibility experiments) has been found in the compounds shown in Figs. 60 and 61. Similar $\rho(T)$ curves are displayed by Np and Pu metal.[18] The inverse magnetic susceptibility of the UX_3 compounds (Fig. 62) generally follows a Curie–Weiss law above the critical temperature T_s and either flattens out or develops curvature as T is lowered below T_s.[144]

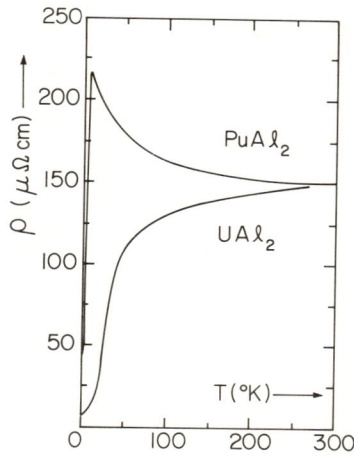

Fig. 60. The electrical resistivity ρ vs. T in UAl_2 and $PuAl_2$, Ref. 18. The steep increases in ρ are thought to result from spin-fluctuation scattering (see text). Note that these compounds do not order magnetically.

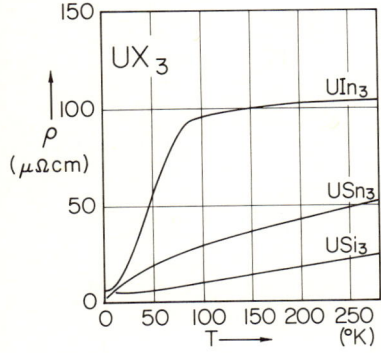

Fig. 61. The electrical resistivity ρ vs. T of three UX_3-type compounds, Ref. 144. These data indicate that UIn_3 is a "spin-fluctuation" compound.

The behavior of these compounds has been correlated with the localized spin fluctuation model[148, 149] in which above T_s the resistivity is dominated by scattering of conduction electrons from the spins of mostly localized or "virtually bound" (see Chapter 3) $5f$ electrons. Arko et al.[18] suggest that below T_s the $5f$ electrons become more itinerant with decreasing T and form well-defined bands. The localized spin fluctuation model predicts a T^2 dependence of ρ below T_s, and this is observed in a number of the actinide intermetallics. The $\rho(T)$ and $\chi(T)$ behavior of these materials is also very similar to that observed in a number of rare earth intermetallics such as $CePd_3$, which is believed to be an intermediate valence compound.[277]

2.5.2. AX_2-Type Intermetallics

Particular effort has already been devoted to studying one member of this class of compounds, UAl_2, which is believed[150] to be a substance showing the spin-fluctuation phenomenon (Chapter 3, Section 2.2).

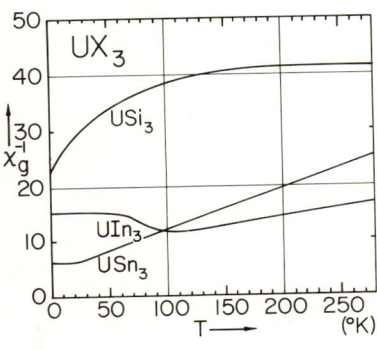

Fig. 62. The inverse gram-magnetic susceptibility χ_g^{-1} vs. T of the three UX_3-type compounds, Ref. 144. Note the absence of magnetic ordering anomalies or divergences as $T \to 0$. These data are typical of spin-fluctuation compounds.

Table 6. Intermetallic Actinide Compounds of the Type AX_2 (A: Actinide), Mostly with Cubic Structure (ρ_m: estimate of magnetic part of resistivity; ρ_{300}: resistivity at 300 K; other symbols are defined in Table 3, p. 14.)

Compound AX_2	a_0 (Å) or structure	T_N (K)	T_C (K)	θ_p (K)	μ_0 (μ_B)	μ_p (μ_B)	ρ_m ($\mu\Omega$cm)	ρ_{300} ($\mu\Omega$cm)	Refs.
UAl$_2$	7.795	—	—	−250	—	3.2	—	150	18, 144, 145, 250
UFe$_2$	7.058	—	172	170	0.38 (Fe); .03 (U)	1.02 (Fe)	60	125	23, 161, 162, 27
UGa$_2$	$c = 4.02$, $a = 4.22$	—	133	133	1.8	3.2	—	325	163, 170, 144
UMn$_2$	7.163	260	—	−2000	—	6.6	64	85	161, 7
UGe$_2$	ZrSi$_2$	—	52	−89	0.80	3.03	2100	—	23
UHg$_2$	$a = 4.98$, $c = 3.22$	70	—	−200	—	3.2	—	—	163, 170
NpAl$_2$	7.785	—	61	—	1.6	2.3	—	≈200	23, 155
NpNi$_2$	7.098	—	32	20	1.2 (Np)	2.3	—	—	23, 249, 7
NpFe$_2$	7.144	—	500	—	1.0 (Np)	—	—	—	249
NpIr$_2$	7.509	7.5	—	−120	0.6	3.1	—	—	23, 155
NpCo$_2$	7.043	≈15	—	−160	0.55 (Np)	3.4	—	—	23, 249, 7
PuAl$_2$	7.840	—	—	−150	—	1.0	—	150	23, 18
PuRu$_2$	7.474	—	—	—	—	—	—	112	146, 147
PuRh$_2$	7.488	—	—	−49	—	0.88	—	117	146, 147
PuIr$_2$	7.531	—	—	—	—	—	—	80	146, 147
PuPt$_2$	7.652	—	6	6	weak FM	0.89	80	113	146, 147

2.5. INTERMETALLIC ACTINIDE COMPOUNDS

Table 7. Intermetallic Actinide Compounds of the Type AX_3. (Symbols as in Table 6, p. 64.)

Compound AX_3	a_0 (Å)	T_N (K)	T_C (K)	θ_p (K)	μ_0 (μ_B)	μ_p (μ_B)	ρ_m ($\mu\Omega$ cm)	ρ_{300K} ($\mu\Omega$ cm)	Refs.
UAl$_3$	4.287			−600		3.4		205	21, 144
UGa$_3$	4.256	70			0.82			130	144, 163, 170
UIn$_3$	4.606	≈ 95 (?)			1.0			105	144, 163, 170
USi$_3$	4.04							25	23, 144
UGe$_3$	4.20							30	23, 144
USn$_3$	4.63			−50		2.5		55	23, 144, 7
β-UH$_3$	6.631		181	176	1.18	≈ 2.6			168, 169
β-UD$_3$	6.621		178	175	≈ 1.2	2.43			168, 169
α-UH$_3$	4.160		178	174	0.9	2.8			168, 169
UTl$_3$	4.688	≈ 80 (?)			1.6			700	163, 170
PuRh$_3$	4.008	6.6		−63		1.0	13.8	55	146
PuPd$_3$	4.102	24		−34	0.8	1.0	131.3	125	146, 7
PuPt$_3$	4.103	40		−36		1.3	77.3	90	146
NpSn$_3$	4.627	9.5				0.28	125	178	166

In UAl$_2$ inelastic neutron scattering experiments do not reveal any transitions between crystal-field states.[150] A very broad ($\Gamma = 50$ meV) line, of magnetic origin, is observed.

Magnetic susceptibility,[151] (also under pressure[152]) specific heat, electrical[18,154] and thermal resistivities,[153] as well as thermoelectric power[153] of UAl$_2$ have been measured. The high value $\gamma = 340$ mJ/mole K^2 [278] of the electronic specific heat coefficient is compatible with the narrow $5f$ band at the Fermi level seen in photoemission experiments.[279] However, no overall consistent picture has yet emerged (probably due to sample quality differences[153]). For instance, the critical spin-fluctuation temperature T_{sf} variously obtained fluctuates between $T_{sf} = 52$ K [151] and $T_{sf} = 4-7$ K.[153] Further efforts appear to be needed to clarify the role of paramagnon or Fermi-liquid contributions to explain the data.

In the series of cubic intermetallics of formula **NpX$_2$** (X = Al, Co, Ni, Fe, Ir), the temperature dependence of the hyperfine fields can rather closely be fitted by an $S = 1/2$ molecular-field theory curve.[154] From this the authors[154] conclude that the ground state of Np behaves as an isolated Kramers doublet.

The same series NpX$_2$, where X = Al, Os, Ir, or Ru, shows a gradual change from localized to itinerant magnetic behavior as the actinide–actinide lattice separation d decreases.[154,155] For example, the ferromagnetic ordered moment, μ_0, as determined from magnetization measurements decreases from $1.6\mu_B$ per Np ion in **NpAl$_2$** ($d = 3.37$ Å) to $0.4\mu_B$ in **NpOs$_2$** ($d = 3.26$ Å). In particular, specific heat and susceptibility measurements[156] on **NpIr$_2$** confirm that it is an itinerant antiferromagnet (as NpSn$_3$, see later) with $T_N = 7.5$ K and a magnetic hyperfine field corresponding to a $\mu_0 = 0.6\mu_B$.[155] A small value of the ordered moment is characteristic of an itinerant ferromagnet, so the evidence indicates that the $5f$ electrons become more delocalized as the lattice constant decreases and the $5f$ wavefunctions overlap more with the $5f$ and $6d$ wavefunctions on neighboring sites. This view is consistent with the rapid reduction in the ordered moment and Curie temperature of NpOs$_2$ as an external pressure is applied.[280] Along the NpX$_2$ series, the magnitude of the isomer shift δ determined by Mössbauer resonance increases by 15 mm/sec (see Fig. 63) which, as explained in Section 2.1.4., is also evidence for delocalization of $5f$ electrons.[154,155] On the other hand, the internal (hyperfine) field H_i determined by Mössbauer resonance is found to be related to μ_0 as follows:

$$H_i = (1.9 \times 10^3 \text{kOe}/\mu_B)\mu_0.$$

This empirical relation is approximately valid for all Np compounds investigated.[90] Such a relationship is expected only for a model of fully localized $5f$

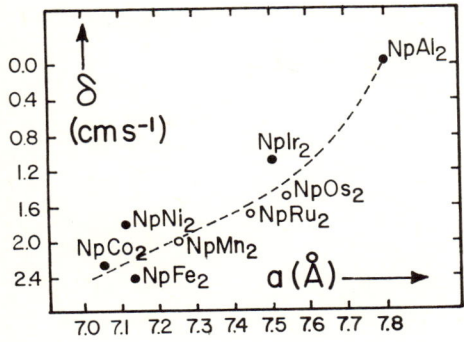

Fig. 63. The Mössbauer isomer shift δ vs. the lattice constant a of NpX_2 compounds, Refs. 154, 155. The decrease in δ results from increasing localization of f electrons with increasing a.

electrons.[157] Thus the neptunium intermetallics exhibit behavior intermediate to that predicted by the band or localized models.

Mössbauer experiments[424] also show a strong dependence of the magnetic ordering of $NpX_{2+\epsilon}$ compounds upon the composition parameter ϵ. For $\epsilon \geqslant 0.02$ and X = Co, Os, or Ir the magnetic ordering no longer exists. These Np-deficient samples have smaller lattice spacings in comparison to the case $\epsilon = 0.0$, and thus the parameter ϵ is equivalent to an external pressure[280] in destroying the magnetic ordering.

The members of series AFe_2 (A = U, Np, Pu, Am) are of the cubic C15-type Laves structure. Some of their properties are summarized in Table 6.[158]

UFe_2 is ferromagnetic, with a Curie temperature in the range of 147–195 K.[159,160,162] (The variation is due to nonstoichiometry,[160] mainly excess Fe.) Polarized neutron diffraction measurements give (reduced to $T = 0$ K) $\mu_{Fe} = 0.44^{27}$–$0.59^{159} \mu_B$ and $\mu_U = 0.03^{27}$–$0.06^{159} \mu_B$. The easy axis of magnetization is the $\langle 111 \rangle$ direction. The magnetization vs. temperature (Fig. 64) curve falls significantly below the $S = 1/2$ Brillouin curve,[162] but can be well fitted by a sum of spin wave and Stoner excitation contributions.[160] The author[160] suggests that UFe_2 behaves like an itinerant ferromagnet. This suggestion is further supported by photoemission data[425] on UFe_2, which show $5f$ spectra closely resembling those of d-electron transition metals. The experimenters conclude that the $5f$ electrons are itinerant in this compound. The very small U–U atomic spacing of $\cong 2.5$ Å leads one to expect more overlap of $5f$ wavefunctions on neighboring sites, leading to the formation of $5f$ bands.

The compounds $NpFe_2$[158,159] ($T_C \cong 500$ K) and $PuFe_2$[158,159] ($T_C \cong 600$ K) show similar behavior to each other (Fig. 64), with local moments of $1.09 \mu_B$ and $0.45 \mu_B$, respectively. There are discrepancies between the bulk moment, and the sum of A and Fe moments per formula unit deduced from neutron diffraction – the latter being $\cong 1 \mu_B$ larger. There is also a large magnetic aniso-

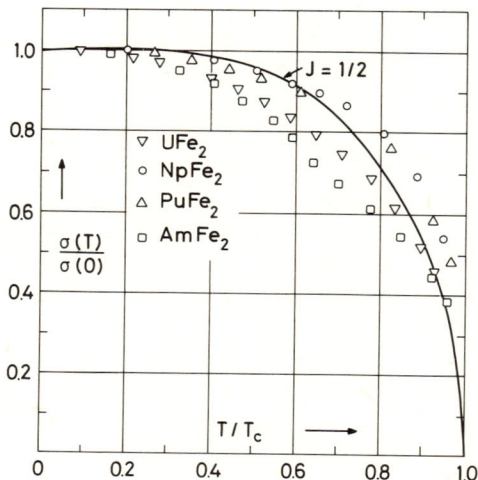

Fig. 64. Reduced magnetization $\sigma(T)/\sigma(0)$ vs. reduced temperature T/T_c for four ferromagnetic AFe$_2$-type intermetallic compounds. For comparison, the $J = 1/2$ Brillouin function is also plotted. From Ref. 158.

tropy in contrast to UFe$_2$ and AmFe$_2$. The authors[159] suggest that a localized electron model is appropriate, with $5f^4$ and $5f^5$ configuration, respectively, but this does not explain the discrepancies between bulk and neutron-measured moments.

AmFe$_2$[159,158] has been investigated by neutron diffraction, magnetization (Fig. 64), and Mössbauer experiments which indicate a Am^{3+}(f^6) configuration, with a nonmagnetic $^7F_0$ ground state. Nevertheless, a small Am moment of $0.4\,\mu_B$ is found, which is negative with respect to the total magnetization. This is explained by an admixture of the f^7 configuration, which has a moment of $7\,\mu_B$. Hence, AmFe$_2$ may be another mixed-valence actinide compound, even though the heavy actinides do not tend to show mixed valency.

Data have also been obtained on **UGa$_2$**,[144,163] **UMn$_2$**[161] (of cubic CaF$_2$-type), **NpH$_2$**, and **PuH$_2$** (probably antiferromagnetic, $T_N \cong 30\,K$) of the AX$_2$ class of compounds.[143,161,163,164] The susceptibility of **PuRh$_2$** (Fig. 65) rises steeply near $T = 10\,K$, indicating a tendency to ferromagnetism, but starts to level off below 5 K.[146,147]

2.5. INTERMETALLIC ACTINIDE COMPOUNDS

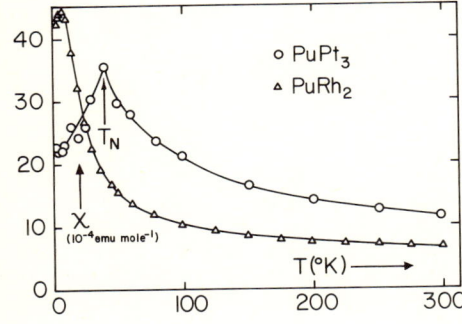

Fig. 65. The molar magnetic susceptibility χ vs. T in $PuPt_3$ and $PuRh_2$, Ref. 146. T_N = Néel temperature. The data clearly indicate antiferromagnetic ordering in $PuPt_3$.

2.5.3. AX_3-Type Intermetallics

UPd_3 with hexagonal structure and D_{6h}^4-$P6_3/mmc$ space group does not order magnetically down to 1 K. In annealed crystals, however, a sharp phase transition is observed at $T_0' = 6$–7 K (depending on sample).[165] This is indicated by a Lambda-type spike in the specific heat (see Fig. 66), a change in the c/a lattice constant ratio by 0.4%, and a kink in the susceptibility vs. temperature diagrams (Fig. 67). Most significant is an increase in the electrical resistivity of a

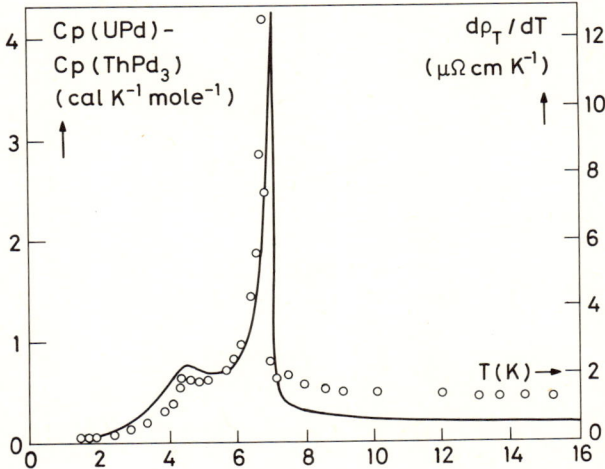

Fig. 66. Specific heat c_p of a single crystal of UPd_3 (with specific heat of $ThPd_3$ subtracted) vs. temperature T (open circles and left scale). Derivative of the resistivity with respect to the temperature $d\rho_T/dT$ vs. T, measured perpendicular to the hexagonal axis (solid curve, right scale). From Ref. 165.

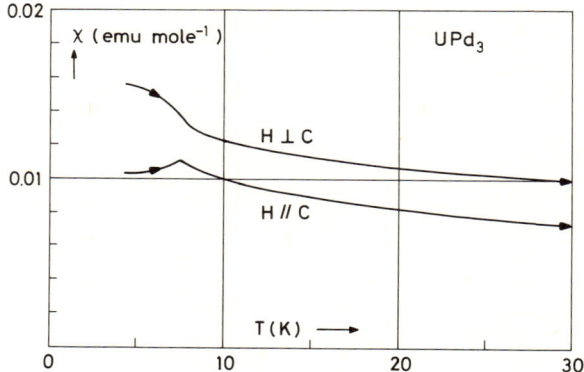

Fig. 67. Magnetic susceptibility χ of a single crystal of UPd$_3$ vs. temperature T, measured in a field of 1 kOe oriented parallel (H ∥ c) and normal (H ⊥ c) to the hexagonal crystal axis. From Ref. 165.

single crystal specimen. For instance, with current flowing normal to the c axis a tenfold resistivity increase is observed when heating from 2 K to 8 K. In a field of 50 kOe, the resistivity ρ below T_0' is increased by a factor of $\cong 5$, whereas above T_0' the field induced an increase in ρ by only a few percent (see Fig. 68). Recent neutron experiments[265] indicate that the apparently first-order phase

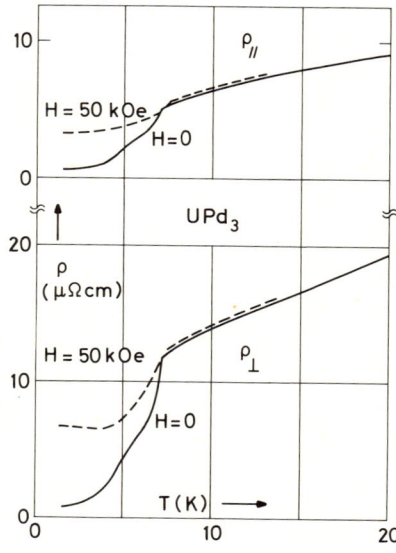

Fig. 68. Resistivity of a single crystal of UPd$_3$ parallel ($\rho_\parallel$) and perpendicular ($\rho_\perp$) to the hexagonal crystal axis vs. temperature T. Solid lines: zero magnetic field H. Broken lines: $H = 50$ kOe applied normal to the axis. From Ref. 165.

transition at T_0' results from a structure change involving a doubling of the unit cell in the a direction. The same experiments led to a model in which the crystal-field ground states of the U^{4+} $5f^2$ ions on the two inequivalent sites in UPd$_3$ are nonmagnetic sublevels of the $^3H_4$ multiplet. The observed spin excitations were also fitted with this model.[265]

A further point is the small ($\gamma = 5$ mJ mole K^{-1}) linear specific-heat coefficient[165] which seems to rule out a narrow $5f$ band and thereby the presence of spin fluctuations (see Part 3.2.2.) to explain the absence of ordering and the resistivity anomaly.

In an earlier independent elastic neutron scattering study[166] on powder samples above 9 K two strong neutron energy-loss lines were observed, which were interpreted[166] as crystal-field transitions from the ground state and from a state at 24 cm^{-1} to a state at 113 cm^{-1}. The lines gradually disappear when the temperature approaches $T = 110$ K. The form factor corresponds well to U^{4+} as found in UO$_2$, and in ThPd$_3$ the lines are absent, which proves that they are related to the presence of a $5f$ electron (Th has none).

Cubic **NpPd$_3$** is antiferromagnetic below $T_N = 55$ K, and the sublattice magnetization $\mu = 2.0\,\mu_B$ is observed to be constant up to 45 K. The transition at T_N is possibly first order and is associated with sharp jumps in the susceptibility and resistivity (see Fig. 69).

NpSn$_3$, which is of cubic AuCu$_3$ structure, also orders antiferromagnetically

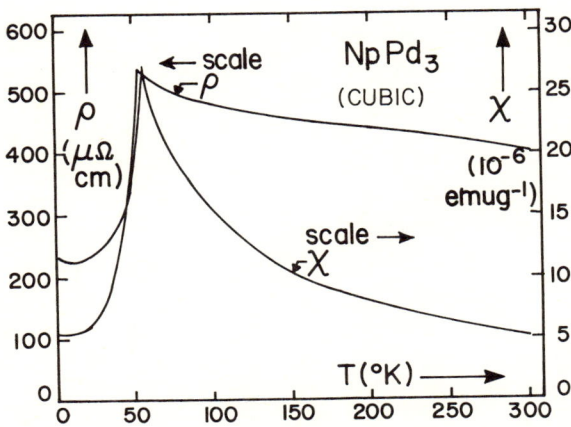

Fig. 69. The electrical resistivity ρ and the gram-magnetic susceptibility χ vs. T in NpPd$_3$, after Ref. 57. The sharpness of the χ anomaly is unusual. The ρ data display spin-disorder scattering.

at $T_N = 9.5$ K with a small moment $\mu = 0.3\,\mu_B$ on the Np ion, as established by Mössbauer resonance on both the ^{237}Np and ^{119}Sn ions.[167] Electronic specific-heat measurements[167] around T_N show a normal paramagnetic behavior above $T = 15$ K, a peak at T_N, and an approximately exponential drop below T_N. The authors[167] point out that this represents the first verification of itinerant f electron antiferromagnetism in any system.

The band electronic specific heat coefficient in the equation $C_E = \gamma_p T$ is, in the paramagnetic state, $\gamma_p = 578 \times 10^{-4}$ cal/mole K^2, which is one of the largest yet observed in an actinide compound. It is a sign of unusually high density of states at the Fermi level in NpSn$_3$.

Hill[298] has established an empirical rule, according to which in uranium intermetallics the $5f$ electrons are localized, if the actinide interionic distance exceeds $\cong 3.5$ Å. Since in NpSn$_3$ the separation between Np ions is 4.9 Å, but the $5f$ electrons are itinerant, the rule clearly does not apply to this Np compound. On the other hand, NpPd$_3$ with an interactinide spacing of 4.0 Å has localized $5f$ electrons.

Plutonium compounds of the **PuX$_3$** type (Table 7) are antiferromagnetic and apparently contain well localized $5f$ electrons.[146] The magnetic susceptibility $\chi(T)$ (Fig. 65) of **PuPt$_3$** reveals a sharp peak at $T_N = 40$ K and a minimum at $T \cong 20$ K followed by a steep rise as T decreases below 20 K.[146,147]

Data are also available on the hexagonal **PuH$_3$**,[168,169] which is ferromagnetic below $T_C = 101$ K, as well as UD$_3$,[168] UGa$_3$,[170] UIn$_3$,[170] and UTl$_3$.[170]

2.5.4. Miscellaneous Intermetallics

The intermetallic compound **U$_{0.5}$Pt$_{0.5}$**, which has the orthorhombic CrB lattice, is a weak ferromagnet with a saturation moment of $\mu_0 = 0.225\,\mu_B$ per U ion.[157] The saturation moment μ_0 decreases continuously as a function of an externally applied pressure P, reaching a value $\mu_0 \cong 0.01\,\mu_B$ at $P = 20$ kbar.[157] The very surprising observation is that the Curie temperature $T_C = 30$ K remains constant over the same pressure range! In all other cases known to the authors, the application of pressure large enough to change the magnetic moment of a ferromagnet also changes T_C significantly. The decrease of μ_0 with P may result from a pressure-induced increase in the delocalization or bandwidth of the $5f$ electrons, but this idea does not explain why T_C remains constant.

From the paramagnetic susceptibility,[171] one deduces $\mu_\rho = 2.62\,\mu_B$, which is close to the Hund's rule value of $\mu_B = 2.54$ for $5f^1$ or $5f^4$ configurations. Heat capacity measurements yield a magnetic contribution to the entropy $\Delta S \cong (\text{mag})k_B \ln 2$, characteristic of a doublet ground state (hence excluding

$5f^4$), and a variation of C(mag), between 4 and 40 K, indicative of ferromagnetic spin waves.[172] Hence, in UPt one may envisage a crystal-field split $J = 5/2$ state, giving a low moment, or a borderline case between localized and itinerant ferromagnetism. The X-ray photoelectron spectroscopy results[281] showed that the position, shape, and intensity of the $5f$ spectrum in UPt are the same as in α-uranium and thus support an itinerant model for the f electrons. Neutron diffraction, resistivity, and specific heat measurements under pressure are needed to clarify the behavior of UPt.

The orthorhombic compound **UPt$_2$** is ferromagnetic below $T_C = 10$ K, and the saturation moment is only $0.1\,\mu_B$.[282] The XPS spectrum is similar to that of UPt.[282]

UCu$_5$[173] with the AuBe$_5$(C15$_b$) structure has the space group $F\bar{4}3m$, and has an effective magnetic moment of $\mu_p = 3.46\,\mu_B$ and $\theta_p = -260$ K, as deduced from the measurement of the susceptibility, which may be described by a Curie–Weiss law between 100 and 800 K (see Fig. 121). This moment is close to the one expected for U^{3+} in the $^4I_{9/2}$ state.

The Knight shift of the Cu^{63} nuclear absorption spectra were also measured for the two inequivalent Cu ions which occur in this structure. These results[174] are somewhat different from earlier data[175] which yielded $\mu_p = 2.3\,\mu_B$, $\theta_p = -78$ K, and which led the authors to postulate a $5f^4$ configuration for U.

An antiferromagnetic phase below $T_N = 15$ K has also been observed.[173] These authors postulate the $5f^2$ configuration for the compound, and indicate the possibility of a mixed valence for uranium in **UNi$_{5-x}$Cu$_x$**.

A mixed valence theory[52] for UNi$_{5-x}$Cu$_x$ has been proposed by Razafimandimby and Erdös (see p. 168).[443] In particular, the peculiar increase of the lattice constant for $x > 4$ is attributed to a partial decrease in U valency.

X-ray-excited photoelectron spectroscopy data[299] indicate that in **UNi$_5$** the configuration is $5f^3$, and in UCu$_5$ and UNi$_{0.5}$Cu$_{4.5}$ a mixed valence configuration with fewer $5f$ electrons than in UNi$_5$ is present.

Nd$_{1-x}$U$_x$Co$_5$ is, in the concentration range $0 < x < 0.6$, of the CaCu$_5$ structure (at $x = 0.6$, $c/a = 0.84$). It shows[176] a linear decrease with x of the Curie temperature from 900 K to 600 K and of the molar magnetic moment with x from $10\,\mu_B$ to $7\,\mu_B$. The authors[176] interpret this by assuming that the magnetic moment is entirely due to the Co ions, that the U is nonmagnetic, and that the U electrons gradually fill up the Co $3d^-$ band. In addition, it is observed that the temperature of the spin-reorientation transition, in which the moments spontaneously turn from the hexagonal basal plane into the c axis, decreases from $\cong 250$ K at $x = 0$ to 160 K at $x = 0.4$.

The compounds **UCu$_2$P$_2$** and **UCu$_2$As$_2$** are ferromagnets with Curie temperatures of 217 K and 140 K, respectively.[426] These compounds have the hexagonal layer lattice of the $CaAl_2Si_2$ type, space group $P\bar{3}m1$. The U–U spacing within a layer is $\cong 4$ Å, and the separation between layers is $\cong 6$ Å, leading to the expectation[426] of much larger exchange interactions within layers than between them. Thus, these compounds are of interest as possible nearly two-dimensional ferromagnets.

UAl$_4$ is a typical spin-fluctuation compound (see Fig. 120).[144]

To conclude the topic of intermetallics, a brief comment on *solid solutions* of uranium with other metals is in order. U metal is nonmagnetic, but in its solid solutions U may or may not be magnetic. For instance, U dissolved in Th is magnetic,[177] in Ti it is nonmagnetic[178] (and superconductive). This is usually interpreted[178] by assuming that in Th, which has a larger atomic radius than U, the U atoms are too far apart for a wide $5f$ band to form. At the same time, the energies of uranium $5f$ electrons and the Th $6d$ bands do not match to hybridize. On the other hand, the $5f$ electrons seem to have a compatible relationship with the $3d$ electrons of Ti.

2.6. Actinide Oxides

The insulating oxides of the actinides such as UO_2 and NpO_2 generally have the cubic CaF_2-type lattice (Fig. 70), and only a few are known to be magnetically ordered. Historically, the oxides were the first actinide compounds to be studied, with their thermodynamic properties coming under careful scrutiny.[131, 179–183] The discovery of the first-order phase transition in UO_2 in 1965 stimulated much of the subsequent experimental and theoretical work on actinide compounds. Some of the pertinent data on actinide oxides is summarized in Table 8. X-ray photoemission data will be presented in the Section on NpO_2.

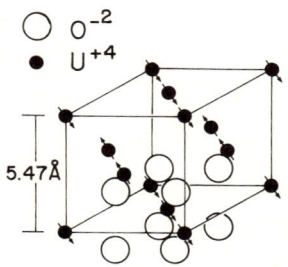

Fig. 70. The unit cell of UO_2, showing the directions of the ordered moments on the U ions (dots) proposed from neutron experiments, Refs. 11, 184. Only one cube of oxygen ions (open circles) is shown.

Table 8. Selected Properties of Actinide Oxides (ΔS: entropy of ordering; other symbols are defined in Table 3, p. 14.)

Compound	Lattice constants	T_N (K)	μ_0 (μ_B)	μ_p (μ_B)	θ (K)	ΔS, $\frac{\text{cal}}{\text{mole-deg}}$	Lattice type	Refs.
UO_2	$a_0 = 5.470$	30.8	1.8	3.2	220	3.0	Cubic	11, 182, 183, 185
U_4O_9	$a_0 = 5.442$						Cubic; $T < 338°$ Rhomb.; $T > 338°$	180, 185, 199
α-U_3O_7	$a = 5.472$ $c = 5.397$						Tet.	179, 185
β-U_3O_7	$a = 5.363$ $c = 5.531$						Tet.	179, 185
U_3O_8	$a = 6.815$ $b = 11.825$ $c = 4.132$						Ortho.; $T < 400$ K (Hex; $T > 400$ K)	179, 185, 198
NpO_2	$a_0 = 5.434$	25(?)	≅0.01	2.95		3.6	Cubic	182, 200, 201, 202, 203
$UFeO_4$	$a = 4.88$ $b = 11.93$ $c = 5.11$	55	4.5	5.42			Ortho.	205, 207

2.6.1. Uranium Dioxide

Uranium dioxide, UO_2, has been by far the most intensively studied actinide oxide. Neutron diffraction[11,184,300] shows that UO_2 becomes antiferromagnetically ordered below $T_N = 30.8$ K, and one possible arrangement of the magnetic moments is shown in Fig. 70. It is known that the moments lie in the planes of the ferromagnetic layers perpendicular to a crystal axis with the moments of each plane antiparallel to those of the neighboring planes. However, the orientation of the magnetization of a layer with respect to the crystal axes in the plane is not certain.[184] The data of Fig. 71 demonstrate that the transition at $T = T_N$ is of the first order, the sublattice magnetization falling from $\cong 50\%$ of its maximum value ($1.8\,\mu_B$ at $T = 0$ K) to zero within an interval $\Delta T/T_N = 0.001$. The first-order nature of the transition is also indicated by the absence of critical scattering[11] near T_N and by the sharp decline of the magnetic susceptibility[185] as T is lowered through T_N (Fig. 72). Below T_N the susceptibility is constant.

The specific heat *vs.* T data[182,406] of Fig. 73 reveal the usual anomaly at T_N. By assuming that the lattice entropy of the isomorphic nonmagnetic ThO_2 is equal to that of UO_2, the magnetic part of the entropy of UO_2 can be obtained by subtraction of the ThO_2 entropy from the total entropy of UO_2.[182] The results shown in Fig. 74 indicate that the maximum magnetic entropy (3.0 cal/deg-mole) is achieved only at high temperatures ($T \cong 300$ K). This may

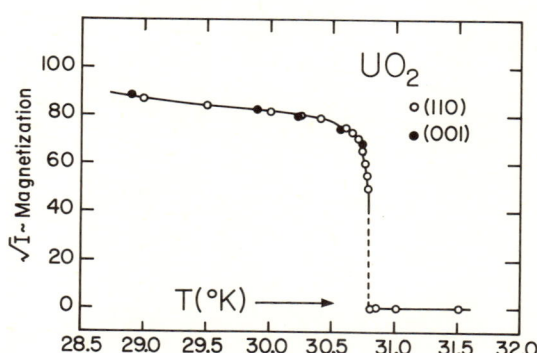

Fig. 71. The square root of the neutron scattering intensity, I (proportional to the sublattice magnetization), *vs.* T of UO_2 for two Bragg reflections having indices ⟨110⟩: open circles and ⟨001⟩: full circles. Ref. 11. Note the first-order transition at $T_N = 30.8$ K, which has been the subject of several theoretical studies (see p. 135).

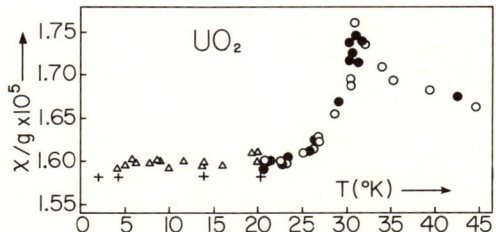

Fig. 72. The gram-magnetic susceptibility χ vs. T of UO_2, from Ref. 185. The data show an anomaly at the Néel temperature 31 K.

be due to the persistence of the short-range order above T_N in the ferromagnetic layers, which is suggested by the large positive Curie–Weiss constant $\theta = 220$ K.[186]

As we shall see later, the question of the magnitude of a volume discontinuity (if any) at T_N is a crucial one for the evaluation of several theoretical models. The data of Fig. 75 indicate a volume change $\Delta V/V \cong 6 \times 10^{-5}$ at $T = T_N$, and discontinuities in the elastic constants of UO_2 were also found in the same experiment.[39, 187]

The thermal conductivity[188, 398] is reduced by two orders of magnitude near $T = T_N$, as Fig. 76 shows. The experimenters attribute this effect to a strong magnon–phonon interaction.

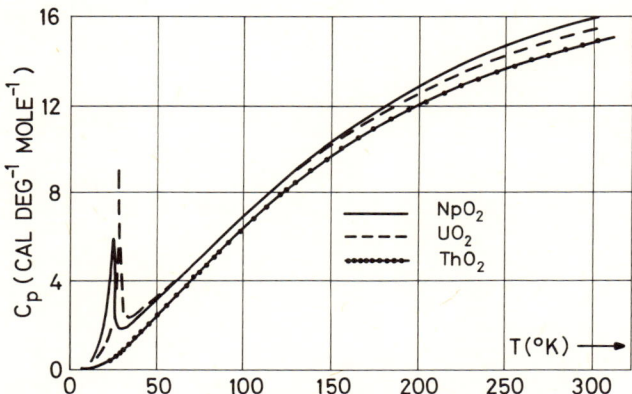

Fig. 73. The molar heat capacity c_p of NpO_2, UO_2, and ThO_2 vs. T, according to the experiments of Refs. 182 and 406. Note that the NpO_2 c_p anomaly near 25 K, even though similar to that in UO_2 near $T_N = 31$ K, results from a quite different type of transition.

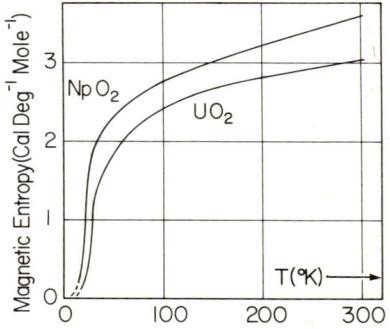

Fig. 74. The magnetic molar entropy of NpO_2 and UO_2 vs. T, as deduced from heat capacity measurements, Refs. 182, 406. Note that the entropy data of NpO_2 are comparable to those of UO_2, despite the observation that the Np ordered moment, if any, is at least a factor of ten smaller than that of the U atom in UO_2.

Another important question concerns the paramagnetic ground state of the U^{4+} ion in UO_2. As will be shown later, theory predicts either a singlet or a triplet. Measurements of the antiferromagnetic resonance spectrum of UO_2 have been interpreted[189] as indicating a singlet ground state, but all other experiments point to the triplet as the state of lowest crystal-field energy. For example, a determination of the spin-wave dispersion branches in single crystals of UO_2 has been carried out with inelastic neutron scattering,[25] and the results were explained by means of a model involving a low-lying triplet with the singlet too far above it energetically to influence the low-temperature properties. The dispersion curves indicate a strong magnon–phonon coupling as already suggested above. This would result from the ability of a lattice distortion (or a "soft" phonon) to remove the orbital degeneracy of the triplet state, producing a split-off level of lower energy. If the level were a singlet, the ion would not interact strongly with phonons, since the singlet could not be further split. Additional evidence for a triplet ground state is provided by a study of the magnetic properties of UO_2–ThO_2 solid solutions, which are discussed in the next section.

The presence of an internal distortion in the ordered state of UO_2 has been verified by neutron diffraction experiments.[184,190] The observed distortion is a

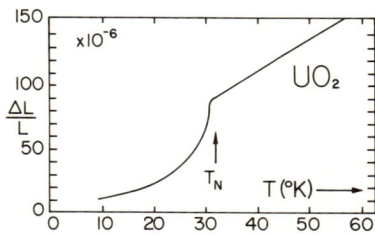

Fig. 75. The relative length change $\Delta L/L$ vs. T of single crystals of the cubic compound UO_2. T_N = Néel temperature, Ref. 187. These data show the presence of a magneto-elastic interaction in UO_2.

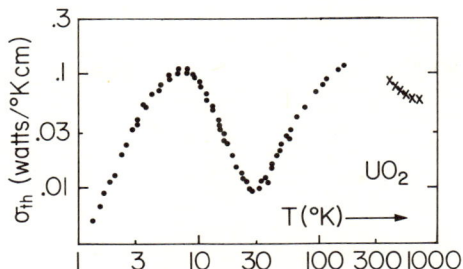

Fig. 76. The thermal conductivity σ_{th} vs. T (logarithmic scale) of UO_2, from Ref. 188. Note the pronounced minimum near $T_N = 31$ K, which thus far has no theoretical explanation.

shear deformation (of magnitude 0.014 Å) of the oxygen sublattice, the position of the uranium ions remaining unchanged (Fig. 102). The distortion is not one of the types considered by the Allen theory[14] and its subsequent modification[283] of the first-order transition in UO_2, but it is in accordance with later theoretical work.[290] (See Section 3.1.4b for a discussion of these theories.)

In a Raman-scattering experiment,[191] two peaks appear at 232 and 275 cm^{-1}. It is known that at these frequencies odd-parity optical phonons exist. The authors[191] could not explain the presence of these peaks, since they are forbidden in Raman scattering, if one considers only $q = 0$ distortions. However, the breaking of the selection rule may be understood on the basis of the lattice distortion. Hence this experiment supports the conclusions of the delicate neutron diffraction study concerning the internal distortions of UO_2.

Optical reflectivity measurements[192] in the photon energy range 0–13 eV have been analyzed in terms of the complex dielectric function. The peaks are thought to derive from two pairs of transitions: from the narrow $5f$ states to the split $6d$ band, and with charge transfer from the $2p$ band of oxygen to the same $6d$ band. The crystal-field splitting of the $6d$ band ($t_{2g} + e_g$) is found to be $10Dq = 2.8$ eV.[192] The optical data from various sources conflict, however, as to the location of the f levels relative to the band.[284]

2.6.2. UO_2–ThO_2 Solid Solutions

UO_2 forms solid solutions with a weakly diamagnetic ThO_2.[193–196] Note that thorium has no $5f$ electrons so that there is no magnetic moment per Th ion. Fig. 77 shows the variation of the lattice constant a, and Fig. 78b the

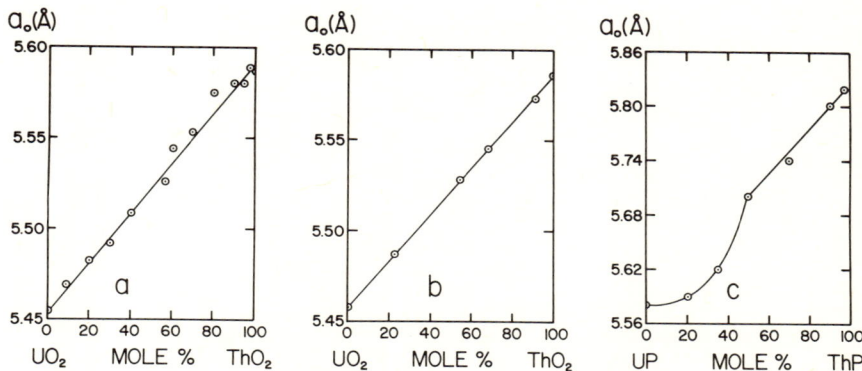

Fig. 77. The lattice constant a_0 vs. the molar percentage of replacement of uranium by thorium in a) UO_2 (Ref. 195); b) UO_2 (Ref. 193); c) UP (Ref. 252). Note the nonlinearity of the data on UP-ThP, where a transition near 40% Th appears to take place.

variation of the Curie–Weiss constant θ and of the magnetic moment μ per U ion.[193] A point of interest is that θ does not extrapolate to zero at the pure ThO_2 limit. This may be the result of "clustering" of U ions in the dilute limit. The same phenomenon was observed in earlier experiments on UO_2-ThO_2,[195] as well as on Up-ThP solid solutions (Fig. 78c).[143]

The most important experiment is the measurement of the temperature dependence of the inverse paramagnetic susceptibility χ^{-1} per mole of UO_2 for

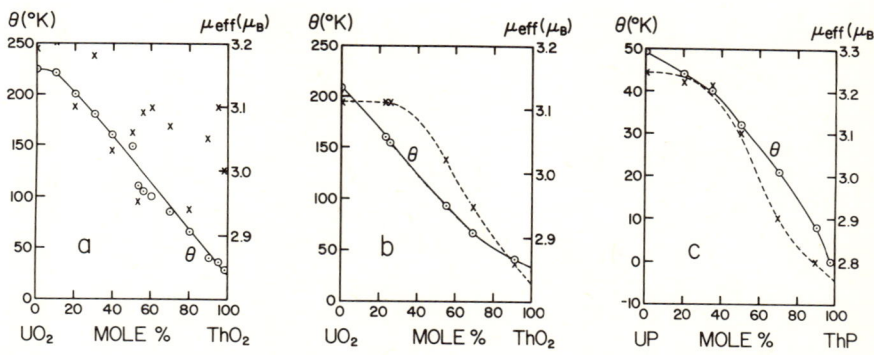

Fig. 78. The paramagnetic Curie constant θ and the effective paramagnetic moment μ_{eff} vs. the molar percentage of replacement of uranium by thorium in a) UO_2 (Ref. 195); b) UO_2 (Ref. 193); c) UP (Ref. 252). Circles and solid curves: θ. Crosses and dashed curves: μ_{eff}. The μ_{eff} data in parts b) and c) are similar, even though the lattice constant data on the same two materials in Fig. 77 are quite different.

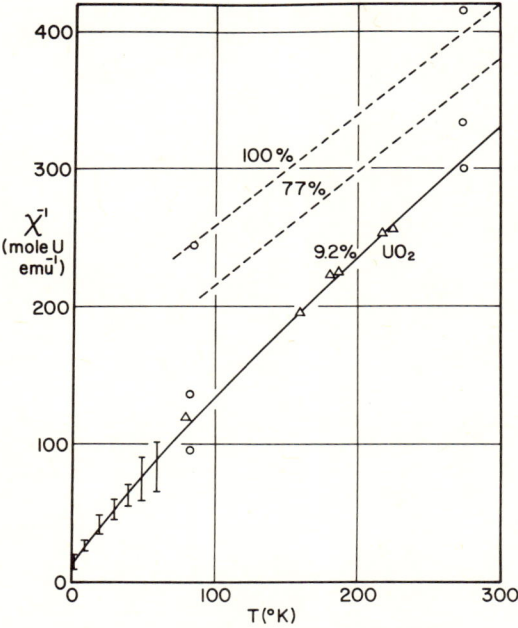

Fig. 79. The inverse molar uranium susceptibility χ^{-1} vs. T of UO_2–ThO_2 mixtures, for three percentages of UO_2. Circles and triangles: data of Ref. 193. Vertical bars: 9.3% UO_2, Ref. 194. Curves: theory based on crystal-field model described in text.

very dilute solutions of UO_2 in ThO_2 (Fig. 79).[193-194] The idea is to "turn off" the complicated magnetic exchange interactions between U^{4+} ions by dilution, while maintaining (hopefully) the same crystal-field environment by choosing isostructural ThO_2 as the solvent. Thus, one expects the susceptibility to reflect the crystal-field levels unperturbed by exchange, even down to low temperature. As may be shown using the formula given for χ in the section on crystal-field theory, in the case of a singlet ground state, $\chi^{-1}(T)$ should increase rapidly as T is lowered to the point where $k_B T$ becomes equal to the separation between the singlet and the next state in energy. Since no such effect is observed in the data of Fig. 79, the experiments are strong evidence for a triplet ground state in UO_2.

2.6.3. Other Uranium Oxides

The *oxides of uranium* U_xO_y exist for a continuous range of x and y values,[185,197] but the magnetic and thermodynamic properties of the

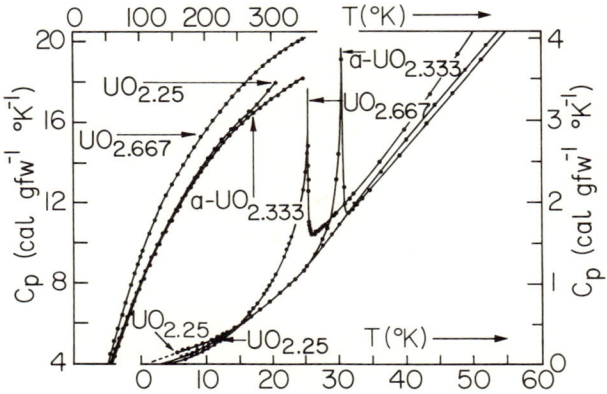

Fig. 80. The heat capacity c_p vs. T of UO_x compounds. The upper left part of the figure shows high-temperature data and low-temperature data are shown in the lower right part of the drawing, Ref. 179; gfw = gram formula weight.

nonstoichiometric oxides are dominated by lattice vacancies and by the presence of interstitial oxygen.[181,197-199] The inverse susceptibilities of UO_{2+x} show anomalies indicative of UO_2-type magnetic ordering at $\cong 30$ K from $x = 0$ to $x = 0.07$,[197] and a sharp peak in the $\chi(T)$ curve at $T = 6$ K is found for $x = 0.1$, $x = 0.33(U_3O_7)$, $x = 0.5(U_4O_9)$, and $x = 0.67(U_3O_8)$.[185] The origin of these peaks is still a mystery, because corresponding anomalies in the specific heats of these materials were not found at this temperature. There is a peak in the specific heat of U_4O_9 at $T \cong 300$ K, which apparently results from a rhombohedral distortion.[199] The compound U_3O_8 has an anomaly in its specific heat at $T \cong 25.3$ K[179] (Fig. 80), which corresponds to a very weak maximum in $\chi(T)$[185] at this temperature. As T is increased to 400 K, U_3O_8 undergoes a transition to a hexagonal phase.[198]

The situation is complicated by the presence of several structure modifications for the oxides.[179,198,199] For example, U_3O_7 is found in two phases: the α-phase, which is a tetragonally distorted UO_2-type lattice with $c/a = 0.986$; and the β-phase, for which $c/a = 1.031$.[179] There is an anomaly in the specific heat[179] of α-U_3O_7 at $T = 30.5$ K (Fig. 80). U_4O_9 undergoes a cubic to rhombohedral transformation at $T = 338$ K, as X-ray studies show.[199]

Clearly, neutron diffraction experiments are needed to clarify the nature of the low-temperature transitions in the uranium oxides.

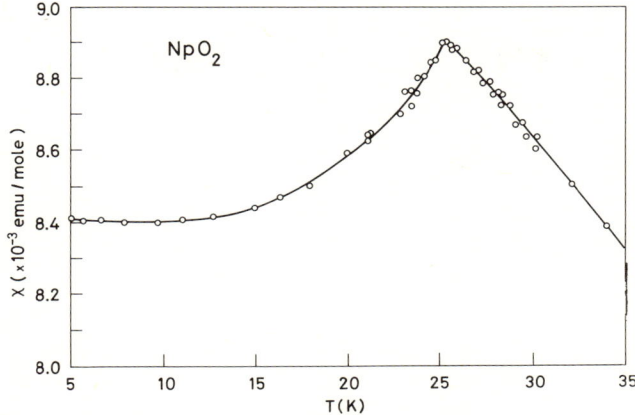

Fig. 81. Magnetic susceptibility χ of single crystals of NpO$_2$ in the neighborhood and below the phase transition point, as a function of temperature T. From Ref. 286.

2.6.4. Neptunium Dioxide

Neptunium dioxide, **NpO$_2$**, has the same crystal structure as UO$_2$ and displays a very similar specific-heat anomaly at $T = 25$ K[182] (Fig. 73), where a peak in the magnetic susceptibility is also observed[200, 286] (Fig. 81). Assuming that the transition at $T = 25$ K is magnetic, one finds that the increase in the magnetic part of the entropy (Fig. 74) exceeds that of UO$_2$. It is surprising that two neutron diffraction experiments[201, 202] failed to detect any measurable ordered magnetic dipole moment above 4.6 K, thereby putting an upper limit of $0.4\,\mu_B$ on the ordered moment per Np ion. Mössbauer resonance experiments[203] found an anomalously low hyperfine field of 50 kG and a temperature-dependent broadening of the resonance, which the experimenters interpreted as an unresolved magnetic splitting due to a very small ordered moment, estimated as $0.01\,\mu_B$. The temperature dependence of the Mössbauer linewidth[203] is shown in Fig. 82 for two samples. Both sets of data resemble typical magnetization vs. T curves, with a well-defined transition at $T = 25$ K. Note the sharpness of the transition in the case of the NpO$_{1.92}$ sample. The smallness of the ordered moment is very surprising in view of the large paramagnetic moment $\mu_p = 3.0\,\mu_B$,[200] the size of the magnetic contribution to the entropy and specific heat,[182] and the transition temperature $T = 25$ K (close to that of UO$_2$).

Thus, NpO$_2$ shows all the signs of a magnetically ordered state below

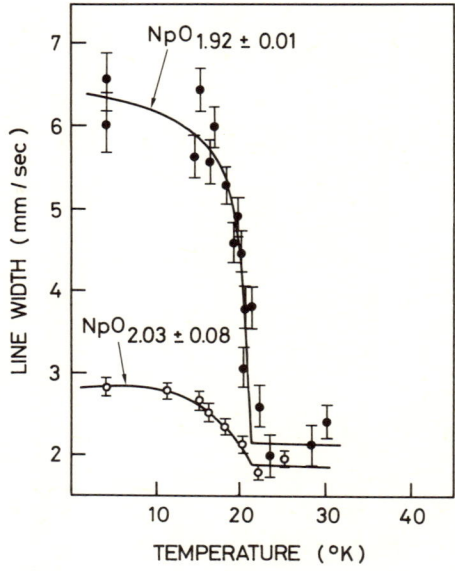

Fig. 82. The line width of the Mössbauer resonance *vs.* T in two non-stoichiometric NpO_2 samples, Ref. 203. The line width may be an unresolved hyperfine splitting resulting from an unusually small ordered moment of $\cong 0.01\,\mu_B$.

$T = 25$ K but only the wisp of an ordered moment. A possible explanation of this paradox is offered later in this review in Section 3.1.4c on theory.

A polarized neutron experiment on a small NpO_2 single crystal yielded a form factor that could best be fitted by the assumption of a $Np^{4+}(5f^3)$ configuration, and reduced the upper limit on any ordered moment to $\cong 0.2\,\mu_B$.[285]

Resistivity measurements show that NpO_2 is a highly resistive semiconductor with an activation energy of 0.40 eV.[285] The high sample resistance ($8 \times 10^8\,\Omega$ at 200 K) made it impossible to obtain accurate results near the transition temperature.

The X-ray photoemission spectra (Fig. 83) of the dioxides of **Th**, U, Np, **Pu, Am, Cm**, and **Bk** have been measured within 50 eV of the Fermi energy.[404] A three-peak structure centered around -25 eV is apparent in each of the oxides. The peaks are attributed to the actinide $6p_{1/2}$, the oxygen $2s$, and the actinide $6p_{3/2}$ orbitals. Comparison of the ThO_2 ($5f^0$) and UO_2 ($5f$) spectra shows clearly the (absence) presence of localized $5f$ states in the (former) latter. In UO_2, as in all the other oxides listed above, the $5f$ states produce very prominent peaks within a few eV of E_F. This is a most important observation and serves as a basis of much of the theoretical work on actinide compounds. The shift of the peak away from the Fermi level in the heavier oxides indicates the more localized character of the $5f$ electrons in the latter.

ACTINIDE OXIDES

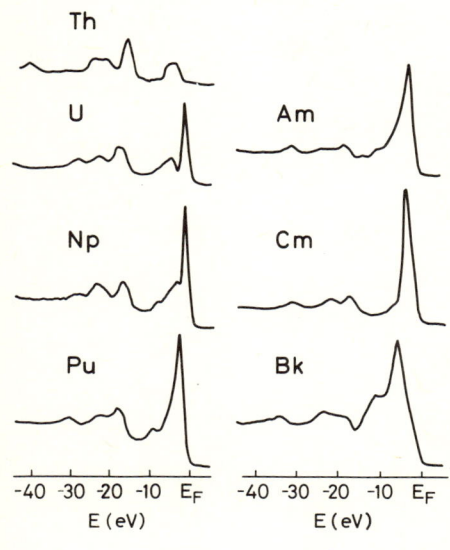

Fig. 83. Photoemission intensity (arbitrary units) vs. electronic energy E for several actinide oxides, Ref. 404. E_F = Fermi energy. The large peaks show the presence of $5f$ electrons near the Fermi level in actinide oxides and their absence in ThO_2.

2.6.5. Intermetallic Oxides

The *intermetallic oxides* $UFeO_4$ and $UCrO_4$ crystallize in the space group $Pbcn(D_{2h}^{14})$ with an orthorhombic unit cell.[204,205] The compound $UFeO_4$ is particularly interesting as a rare example of a ferromagnetic insulator and has been studied by neutron diffraction,[205,206] Mössbauer resonance,[206] and susceptibility measurements.[206] Below $T_N = 55$ K the magnetic moments of the Fe and U ions are antiferromagnetically aligned along the b axis (Fig. 84). At a lower temperature $T' = 42$ K, there is a discontinuous moment jump of ~10%,

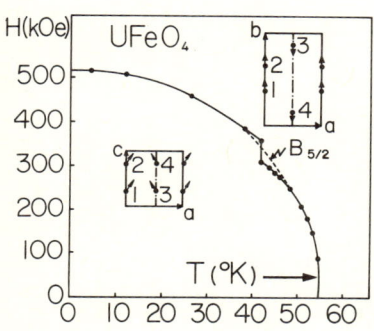

Fig. 84. The internal hyperfine field H vs. T in $UFeO_4$, as deduced from Mössbauer resonance experiments of Ref. 206. The inserts show the directions of the ordered iron moments (arrows) in the ac and ab planes of the orthorhombic unit cell. The field H may be assumed to be proportional to the sublattice magnetization. $B_{5/2}$ is the theoretical Brillouin-curve expected for spin 5/2. Note the spin-canting transition at $T = 42$ K.

as is indicated by the hyperfine field *vs.* T data shown in Fig. 84. This transition is accompanied by a canting of the spins into the ac plane, with a ferromagnetic component $4\mu_B$ per Fe ion in the c-direction and a ferromagnetic component of $2\mu_B$ per Fe ion and $0.4\mu_B$ per U ion in the a-direction at $T = 4.2$ K. The magnetic susceptibility shows two discontinuities at $T_N = 55$ K and $T' = 42$ K.[205] There is a possibility of a small distortion of the lattice at T'.[207]

2.7. Actinide Halides

Data on the magnetic properties of actinide halides are generally incomplete. A variety of lattice types is found, as Table 9 indicates.

2.7.1. Uranium Triiodide

One of the most important examples is *uranium triiodide*, **UI**$_3$, which is orthorhombic with a space group *Cmcm* (see Fig. 85).[12] The specific heat[208] of UI$_3$ is shown in Fig. 86, and an anomaly indicative of a magnetic transition at $T_N = 2.61$ K is seen. The magnetic part of the entropy is seen in Fig. 87, and reaches its full value ($R\ln 2$) only at temperatures much higher than T_N. Extensive short-range order above T_N has been suggested[155] as the explanation for this "missing entropy." The magnetic susceptibility [208,209] shows a broad maximum at $T = 3.2$ K and a marked departure from the Curie–Weiss law in the paramagnetic regime (see Fig. 88) just as is found for **UBr**$_3$.[207,209] The theoretical curves $\chi^{-1}(T)$ shown in Fig. 88 are discussed in the section on theory (Section 3.1.4f).

Table 9. Listing of Actinide Halides (Symbols are defined in Table 3, p. 14.)

Compound	Lattice	T_N (K)	μ_0 (μ_B)	μ_p (μ_B)	θ (K)	Refs.
UI$_3$	Ortho. $a = 13.98$ Å $b = 4.31$ Å $c = 9.99$ Å	2.61	1.98	3.31	5	12, 208, 210, 287
UBr$_3$				3.29	25	210
UCl$_3$		6.5		3.03	−29	210, 427
UF$_4$				3.28	−116	210, 211
UCl$_4$	B.c. tet.			3.29	−62	210, 257
UBr$_4$				3.12	−35	210
NpCl$_4$	B.c. tet.	$T_C = 6.7$		3.08	6.9	213
PaCl$_4$	B.c. tet.	$T_C = 183$		1.04	157	215

2.7. ACTINIDE HALIDES

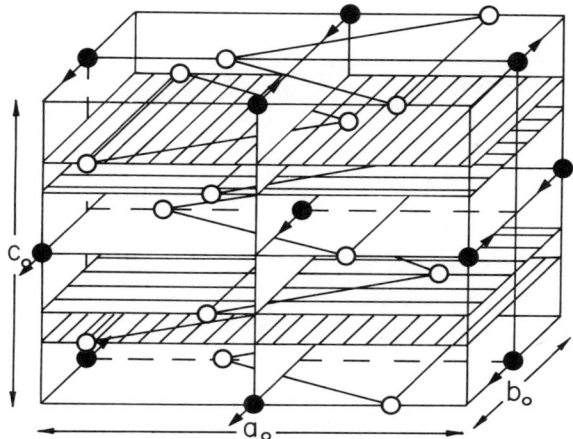

Fig. 85. The unit cell of orthorhombic UI_3. The dots are uranium atoms and the circles are iodine atoms. The arrows indicate the direction of the magnetic moments in the ordered state below $T_N = 2.6$ K, as suggested on the basis of neutron diffraction experiments, Ref. 287.

The most striking result concerning UI_3 is the temperature dependence of the sublattice magnetization as deduced from nuclear magnetic resonance experiments[12] (Fig. 89). The data indicate a first-order transition at $T = T_N$ very similar to that found for UO_2. Neutron diffraction data[287] show an ordered moment of $\mu_0 = 1.98\,\mu_B$ per U ion, but surprisingly indicate a gradual or

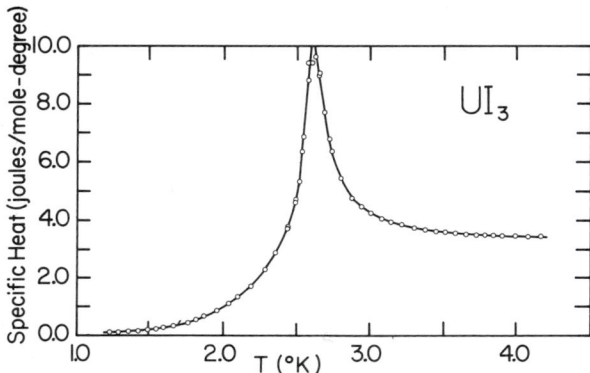

Fig. 86. The molar specific heat c_p vs. T of UI_3, Ref. 208. Note the antiferromagnetic ordering anomaly near $T_N = 2.6$ K. The slow decrease in c_p above T_N may result from strong short-range order or from a complicated crystal-field energy spectrum.

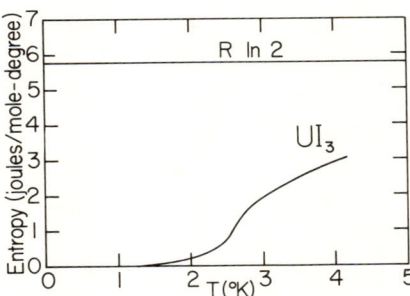

Fig. 87. The entropy *vs.* T of UI_3. The lower curve is deduced from heat capacity measurements, Ref. 208. The horizontal line is the value for a magnetic doublet ground state of the U ion, to which the curve is expected to tend at high temperatures when no long- or short-range magnetic order is present.

second-order transition at T_N (Fig. 90). However, the large error limits on the neutron data and the large short-range order above T_N make the NMR data more reliable in determining the detailed shape of the sublattice magnetization. The magnetic moments are aligned below T_N in ferromagnetic (111) layers with adjacent layers antiparallel. The direction of the moments is parallel to the a axis.[287]

In the cases of the compounds UBr_3 and UCl_3, deviations from Curie–Weiss behavior of the susceptibilities have been observed below 350 K, and the compounds were reported to be antiferromagnetic.[209] However, neutron diffraction in UBr_3 failed to find any long-range three-dimensional ordering.[288] The experimenters suggested that one-dimensional antiferromagnetic ordering of chains of U^{3+} ions could explain the observed susceptibility cusp at $T = 15$ K.[289] The inelastic neutron scattering in UBr_3 detected four transitions, which allowed

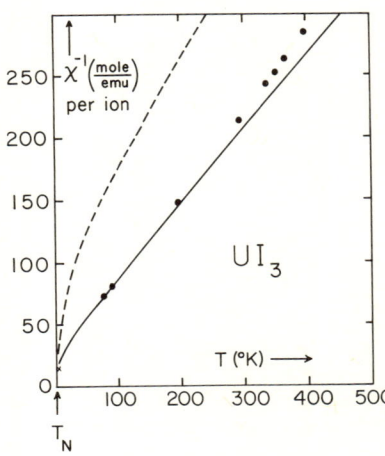

Fig. 88. The inverse magnetic molar susceptibility per U ion χ^{-1} *vs.* T of UI_3. Dots: data of Ref. 209. Cross: data of Ref. 208. Solid and dashed curves: Theoretical predictions of crystal-field models described in the text, Chapter 3.

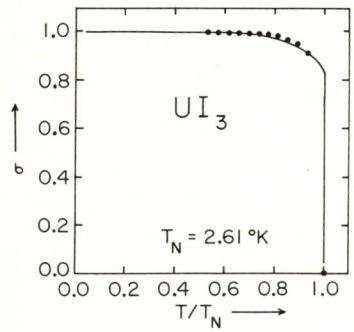

Fig. 89. Relative ordered magnetic moment σ per uranium ion vs. T of UI_3. Dots: NMR data of Ref. 12. Solid curve: Theoretical prediction of the crystal-field model described in the text. Note the first-order magnetic transition, one of the first to be observed in actinide compounds.

the determination of the U^{3+} crystal-field ground state as the $\Gamma_5^{(1)}$ doublet of the $^4I_{9/2}$ multiplet.

Neutron diffraction experiments[427] on UCl_3 revealed a second-order antiferromagnetic phase transition as the sample is cooled down to $T_N = 6.5$ K. Upon further cooling to 3.5 K, however, the intensities of the magnetic neutron reflections begin to decrease rapidly and at 2.5 K vanish completely. This extremely unusual type of transition has not been previously observed in actinide compounds. The experimenters speculate that temperature-dependent displacements of the U and Cl atoms arise because of an assumed magneto-elastic coupling. These displacements would lead to temperature-dependent exchange interactions which might explain the observed behavior. Clearly, much more experimental and theoretical work is needed to clarify the nature of the puzzling transition in UCl_3.

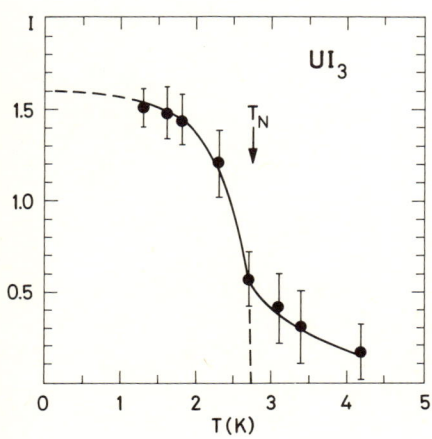

Fig. 90. Temperature dependence of integrated neutron intensity I (in arbitrary units) of the magnetic reflection 131 of UI_3. Note apparent short-range order phenomena above the Néel temperature T_N. The moment is proportional to the square root of the intensity. From Ref. 287.

2.7.2. Other Actinide Halides

Uranium tetrafluoride, **UF$_4$**, does not appear to order magnetically down to 1.3 K,[210] but there is a *Schottky* anomaly (i.e., a broad maximum) in the specific heat *vs. T* data[211] at about 6 K, indicating two closely spaced low-lying crystal-field levels. The magnetic susceptibility[209] is sample-dependent, but generally rises sharply with decreasing T before leveling off and becoming temperature independent below 2 K. These results were explained by a model[210] in which two-thirds of the U^{4+} ions have a nonmagnetic singlet ground state and the remaining U^{4+} ions have a triplet level split into closely spaced singlets by a rhombic distortion of the cube of surrounding F ions. The broadening of the NMR spectrum[212] of ^{19}F in this compound was interpreted as being due to the dipolar interaction between the fluorine nuclear moment and the uranium electronic moment. UF$_4$ is interesting as a possible example of an induced-moment system.[124]

NpCl$_4$ is an example of a ferromagnetic halide. The magnetization as a function of temperature has been determined[213] and falls to zero near $T_C = 6.7$ K more steeply than a Brillouin curve for any spin.

Uranium tetrachloride, **UCl$_4$** (space group $I4_1/amd$), shows a temperature dependence of its magnetic susceptibility which in the region 100–300 K may be fitted well by a Curie–Weiss law. At 19.5 K the susceptibility exhibits a sharp peak typical of an antiferromagnetic transition.[204] Another, broader peak is observed at 10 K. However, neutron diffraction experiments at 4.2 K do not show magnetic ordering.[204]

Also, a similar susceptibility peak is observed in **U$_{0.48}$Th$_{0.52}$Cl$_4$** at 18 K; hence these peaks may not be of magnetic origin.[214] The specific heat c_p does not show any anomaly but rises smoothly in the region 2.4–26 K.[214]

Protactinium tetrachloride, **PaCl$_4$**, is ferromagnetic[215] below $T_C = 183$ K. This Curie temperature is exceptionally high for an ion with only one magnetic electron ($5f^1$). To explain the paramagnetic moment, considerable mixing of the $J = 5/2$ Russell–Saunders ground state with the $J = 7/2$ state has to be admitted.[215]

Since PaCl$_4$ has a slightly larger lattice parameter than NpCl$_4$, one would expect a smaller Curie temperature. To the contrary, it is $T_C(\text{PaCl}_4) = 183$ K, and $T_C(\text{NpCl}_4) = 6.7$ K. It seems that the $5f^1$ configuration is more covalent, i.e., delocalized than $5f^3$, giving rise to large indirect exchange via the chlorine ions.

If it is possible to generalize from the results on U- and Np-tetrachlorides, then once again the increasing localization tendency when progressing along the A series stands out. However, to draw conclusions, on the one hand the results for UCl_4 need to be made more conclusive, and on the other hand, higher A-tetrachlorides would also have to be studied.

The halides CmF_4 and AmF_3 have been investigated by optical absorption and magnetic measurements[428] as examples of compounds with well-localized $5f$ electrons, which should exhibit rare-earth-like behavior. As expected, the paramagnetic Curie temperatures θ are small ($\theta \cong 2\,K$), indicating weak exchange interactions, and the compounds did not order magnetically down to 4.2 K. The surprising results were the magnitudes of the effective paramagnetic moments, $\cong 3\,\mu_B$ and $0.63\,\mu_B$ for CmF_4 and AmF_3, respectively. The actinide ions should be in the $5f^6$ configuration, for which a $J=0$ ($\mu_{eff}=0$) ionic ground state is predicted by Hund's rules. The values of μ_{eff} cannot be explained by thermal population of excited ionic states, because the optical measurements show gaps of 0.41 eV and 0.33 eV, much larger than kT (exper.) in CmF_4 and AmF_3, respectively, separating the ground and first excited states.[428] Clearly, these compounds deserve theoretical investigation.

2.8. Other Compounds

Besides the simple binary compounds, many other higher stoichiometric compounds of the actinides are known. Some of these, for instance the oxides U_4O_9, U_3O_7, and U_3O_8 have already been discussed here (see Section 2.6.3).

Among the miscellaneous binary compounds, U_2S_3 is distinguished by being the only ferrimagnetic actinide compound known.[216]

Two interesting families[216] of ternary compounds derive from uranium sesquinitride, U_2N_3. These are the hexagonal antiferromagnets U_2N_2M, where $M =$ P, As, S, or Se, and the tetragonal ferromagnets U_2N_2Z, where $Z =$ Sb, Bi, or Te. Their properties have not yet been studied in detail.

The absorption spectrum and paramagnetic resonance spectrum of $Cs\,UF_6$ have been studied primarily in order to derive spectroscopic parameters of the U^{5+} ion and its crystal-field parameters in the compound.[217-218]

A class of *semiconducting anisotropic antiferromagnets* has been investigated and includes $CrUS_3$, VUS_3, $CrUSe_3$, $CoUS_3$, and $NiUS_3$. The first three have decentralized, unpaired uranium electrons[219,402] as neutron diffraction results show.

In the complex *americium tricyclopentadienide* $Am(C_5H_5)_3$ the susceptibility data[220] are in agreement with the assumption of an Am^{3+} $J=0$ ion, although Am^{2+} impurities blur the interpretation somewhat. A list of Racah and spin-orbit parameters of the Am ion is given.[220]

The orthorhombic (D_{2h}^{16}-Pnma) **U(OH)$_2$SO$_4$** has been investigated in polycrystalline form. Above 21 K its magnetic susceptibility corresponds well to a paramagnet with a $M_J = \pm 2$ ground-state doublet and a $M_J = \pm 3$ doublet at 139 cm^{-1} of U^{4+}. Below the peak at 21 K, the susceptibility is almost constant: the authors[221,401] exclude an antiferromagnetically-ordered state because their calculated susceptibility decreases more rapidly with temperature than the experimental results. On the other hand, for a Jahn–Teller distortion separating the $|2\rangle \pm |-2\rangle$ states they find the correct susceptibility. For the specific heat the agreement of the Jahn–Teller model with experiment is only within a factor of two.[221] In view of the fact that stoichiometric differences often produce large susceptibility discrepancies (see, for instance, NpC in Fig. 31), neutron diffraction results would be required to confirm the conclusion that magnetic ordering is absent.

2.9. Spectroscopic Data on Actinide Ions

The absorption spectra on actinide ions doped in nonmagnetic crystals such as CaF$_2$[222] or ThBr$_4$[291] or bound in complexes such as the **UCl$_6^{2-}$** complex in [(CH$_3$)$_4$N]$_2$UCl$_6$ have been determined in a number of experiments.[222-234] Typical spectra in the visible and infrared range are shown in Fig. 91a and b for U^{4+} in CaF$_2$. From these spectra[222] and from examining the effect of an applied magnetic field on the absorption peaks, one may infer an energy-level scheme for the actinide ion in the crystal field of the host, as is shown in Fig. 91c. For given values of the spin-orbit interaction ζ, the Slater–Coulomb integrals F_2, F_4, and F_6 for $5f$ electrons, and the parameters of the crystal field (numbering two in sites of cubic symmetry), the energy levels may be calculated by the methods of crystal-field theory as described in Chapter 3. In practice, the interaction parameters listed above are varied to obtain a "best fit" of the observed spectrum of levels. Sets of parameters thus derived are listed in Tables 10 and 11 for various actinide ions.

One sees from Tables 10 and 11 that the spin–orbit interaction in the actinides is about 2000 cm^{-1}, or roughly twice that of the rare earths. A more drastic difference is that of the crystal-field interaction, which is of the same order of magnitude as the spin–orbit term for actinide ions, and an order of magnitude larger than the crystal fields normally found in the lanthanides. As we shall see in Chapter 3, theories of the magnetic properties of actinides require knowledge of at least the spin–orbit and Coulomb parameters, and thus the spectroscopic data are useful in providing these quantities. However, it is

Table 10. D_{2d} Spectroscopic Parameters of U^{4+} (All values are given in cm^{-1})[a]

	ThBr$_4$-U^{4+}	ThCl$_4$-U^{4+}	UCl$_4$
ξ	1777 ± 4	1790 ± 4	1814 ± 5
F^2	42466 ± 64	42752 ± 69	43255 ± 83
F^4	41480 ± 250	40765 ± 314	39908 ± 309
F^6	26773 ± 157	25171 ± 215	25109 ± 214
α	30 ± 1	31 ± 4	30 ± 1
$\beta/12$	−47 ± 3	−38 ± 4	−51 ± 6
γ	999 ± 59	959 ± 69	1067 ± 88
B_0^2	−1062 ± 45	−1047 ± 60	−1032 ± 91
B_0^4	1119 ± 45	1045 ± 159	1845 ± 213
B_4^4	−2360 ± 62	−2972 ± 83	−2817 ± 94
B_0^6	−3285 ± 205	−3190 ± 249	−2521 ± 266
B_4^6	75 ± 140	−187 ± 131	59 ± 242

[a] From Ref. 291. The symbols are defined on pp. 92, 106, 107, and 112.

extremely questionable whether the crystal-field parameters derived from these data correspond to those of any of the actinide compounds discussed in the previous sections. In the case of the NaCl-type metallic actinide compounds, for example, even the signs of the effective charges of the ligands are uncertain.[235] These effective charges determine the crystal-field parameters.

An additional complication stems from coupling of electronic transitions of the actinide $5f$ electrons to the normal modes of vibration of the surrounding ligand ions. This results in a "vibronic" splitting of the pure electronic lines in the absorption spectrum, as shown in Fig. 92 for U^{4+} in Cs_2ZrCl_6 for two energy levels.[225] The centers of the vibronic patterns give the energy levels of the ion. The vibronic splitting assists the identification of the levels, since certain vibrational modes do not couple to a given electronic state, resulting in the absence of the corresponding line in the vibronic pattern (Fig. 92, bottom).

In addition to the absorption of light described above, the resonant absorption of microwave-frequency phonons by actinide ions has been studied[236-237] and is called ultrasonic paramagnetic resonance (UPR). Uranium-doped CaF_2 has been the system most investigated by this method, partly because of the laser properties of U^{3+}: CaF_2 and partly because of the strong spin–lattice interaction of the U^{4+} ion. This latter interaction has been proposed as the crucial factor responsible for the first-order phase transition in UO_2, and the UPR experiments have enabled the spin-lattice coupling coefficients to be deduced for U^{4+} in isomorphic CaF_2.[236] A strong coupling of the order of 100 cm^{-1} (unit strain)$^{-1}$ was found.

In *neptunium hexafluoride*, **NpF$_6$**, optical absorption and magnetic

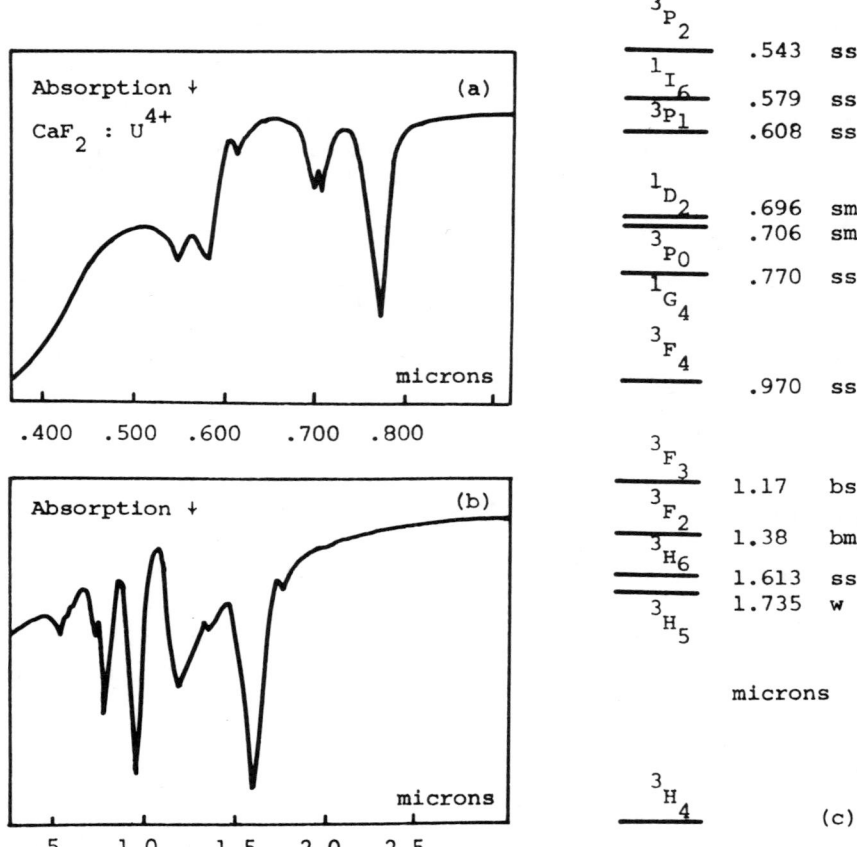

Fig. 91. The optical absorption spectrum of the U^{4+} ion in CaF_2: a). between 0.4 and 0.9 microns; b) between 0.5 and 3.0 microns. In part c), the energy-level scheme of the U^{4+} $(5f^2)$ ion is displayed, as deduced in Ref. 222 from the experimental results. (s: sharp or strong, m: medium, b: broad, w: weak).

measurements, in particular the temperature-independent susceptibility, are difficult to reconcile with theoretical calculations,[238] but lead to the value $\zeta = (2250 \pm 50)\,cm^{-1}$ for the neptunium spin–orbit coupling constant.

The absorption and emission spectra of U^{4+} in $ThBr_4$ single crystals have been analyzed between 0 and 22000 cm^{-1}. Whereas in a first paper[239] an assignment to calculated levels is done, in a second paper[240] the authors express the view that due to the mixing of multiplets with different values of J a unique assignment is presently not possible. Comparing different measurements, one

2.9. SPECTROSCOPIC DATA ON ACTINIDE IONS

Table 11. Crystal-field Parameters (see p. 112; in cm^{-1}) and Mean Radii (in a.u.) of Actinides and the Lanthanide Pm^{3+}, Isoelectronic to Np^{3+} in Similar Surroundings. (Error limits in brackets. The parameters for Am^{3+} were estimated and held constant for all fittings.)

	B_0^2	B_0^4	B_0^6	B_6^6	$\langle r^2 \rangle$	$\langle r^4 \rangle$	$\langle r^6 \rangle$
Np^{3+}:LaBr$_3$[301]	101	−661	−1339	955			
Np^{3+}:LaCl$_3$[301]	165 (26)	−623 (44)	−1615 (48)	1041 (33)			
Pm^{3+}:LaCl$_3$[301]	143 (18)	−395 (29)	−666 (30)	448 (21)			
U^{3+}:LaCl$_3$[303]	260 (64)	−532 (139)	−1438 (113)	1025 (88)	2.36	11.65	107.6
Pu^{3+}:LaCl$_3$[303]	226 (25)	−543 (48)	−1695 (45)	1000 (40)	1.16	3.49	22.19
Am^{3+}:LaCl$_3$[433]	230	−610	−1590	980			
Cm^{3+}:LaCl$_3$[433]	246	−671	−1410	921			

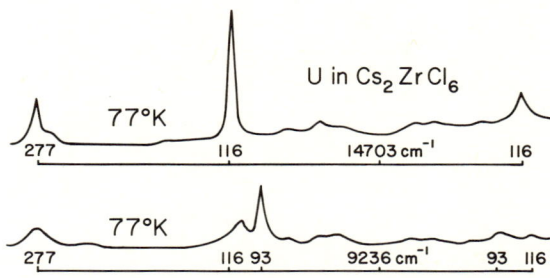

Fig. 92. Illustration of vibronic selection rules helping the identification of the spectrum of U^{4+} in Cs_2ZrCl_6. Top spectrum: Γ_1 level, Γ_{5u} vibration missing. Bottom spectrum: a different electronic level, with Γ_{5u} vibrational peak at 93 cm^{-1}. From Ref. 225.

arrives at the following ranges of the spectroscopic parameters of U^{4+}: $F^2 = 38-44$, $F^4 = 31-42$, $F^6 = 18-31$, $\zeta = 1.66-1.74$ in thousands of cm^{-1}.

The paramagnetic resonance of ^{237}Np^{4+} in single crystals of Cs_2ZrCl_6 (at 1.6 K) shows[241] that the ion has a Γ_8 (quartet) ground state (perhaps split) and occupies a site of O_h symmetry. The Lea, Leask, and Wolf parameter (see Chapter 3) which fits the data best is $x = -0.567$, although in octahedral coordination one would expect $x > 0$. The anisotropy pattern of the nuclear hyperfine splitting is not understood.[241] Results on Np^{4+} in ThO$_2$ corroborate[242,243] the assignment of Γ_8 as the ground state, split into two Kramers doublets.

Racah parameters E_i and spin–orbit coupling constants have also been determined for the **Am**$^{3+}$ ion from AmCl$_3$ in LaCl$_3$, and in fused LiNO$_3$. The values are in the range $E_1 = 3582-4031$ cm^{-1}, $E_2 = 12-21$ cm^{-1}, $E_3 = 334-393$ cm^{-1}, and $\zeta = 2548-2605$ cm^{-1}.[244,245,230]

A very thorough investigation of the energy-level spectrum of **Np**$^{3+}$ in LaCl$_3$ and LaBr$_3$ was carried out by W. T. Carnall et al.,[301] including data taken by S. P. Cook.[302] Np concentrations of 0.01%, 0.1%, and 1% were used. The region 3365–10130 Å was photographed with a 3.4 m Jarrel–Ash instrument of 5 Å/mm reciprocal dispersion. A prism-grating spectrophotometer was used

Table 12. Observed and Calculated Crystal-Field Splittings (in cm^{-1}) of the Ground State of Np^{3+} in LaCl$_3$ (C_{3h} Symmetry)^a

E (obs)	0	125	170	—	(247)	465
E (calc)	−19.8	110.9	173.1	185	233	455.8

^{a} From Ref. 301.

2.9. SPECTROSCOPIC DATA ON ACTINIDE IONS

Table 13. Slater Integrals F^i and Spin–Orbit Coupling Parameter λ for Trivalent Actinides in LaCl$_3$ Host (in cm^{-1}) as Determined from Spectroscpic Data

	F^2	F^4	F^6	λ
U^{3+} [303]	39715 (218)	33537 (302)	23670 (211)	1623 (4)
Np^{3+} [303]	44907 (161)	36918 (245)	25766 (221)	1938 (2)
Pu^{3+} [303]	48670 (154)	39188 (294)	27493 (133)	2241 (2)
Am^{3+} [433]	51800	41440	30050	2580
Cm^{3+} [433]	55109	43803	32610	2903

between 2000 and 25000 Å at 4.77 K and 298 K. The Hg lamp-excited fluorescence was measured with a 21-ft. Paschen-Runge spectrograph; these measurements were supplemented by N$_2$-pumped dye laser excitation experiments. The data of Cook included Zeeman spectroscopy. The observed level structure was fitted by varying first the four crystal-field parameters appropriate to the C$_{3h}$-symmetry, then varying the parameter of the free-ion Hamiltonian (see 3.2) in the numerical diagonalization process of the Hamiltonian matrix. The latter included 250 states. Of these, 45% could be assigned to the observed energy levels in the range of 0–26000 cm^{-1}, with an average error of 20 cm^{-1} per level.

The best fit values of the principal ionic parameters as well as the crystal-field parameters are given in Table 11 and 12.

It is interesting to note that the authors found that the magnetic moment of the ground state is very nearly zero. The composition of the six crystal-field split eigenstates arising from the ground level of Np^{3+} is given in their Table III. There is very little mixing ($\leqslant 1\%$) of J values other than $J = 4$ in the ground manifold.

It is instructive to compare the crystal-field parameters (see Eq. 3.1.2–16) of the $5f^4$ Np^{3+} and the isoelectronic $4f^4$ lanthanide Pm^{3+}. If one scales the Pm^{3+} parameters using the expression

$$B_q^k(\text{Ac}) = [\langle r_{\text{Ac}}^k \rangle / \langle r_{\text{Ln}}^k \rangle] B_q^k(\text{Ln})$$

Table 14. Comparison of Experimental and Theoretical Values of the Slater Integrals F^i and Spin–Orbit Coupling Constant for U^{3+} Ions in LaCl$_3$. (Numbers in cm^{-1}. Error in last figures in brackets.)

U^{3+} : LaCl$_3$	F^2	F^4	F^6	λ
Experiment [303]	39175 (218)	33527 (302)	23670 (211)	1623 (4)
Theory [305]	71442	46370	22918	1898

one obtains parameters which are roughly twice as large as the ones found experimentally. This suggests small covalency in $Np^{3+}:LaCl_3$.

An analysis of U^{3+}, Np^{3+}, and Pu^{3+} in $LaCl_3$ similar to the one discussed above was carried out by H. Crosswhite et al.[303,304] The results concerning the Slater and spin–orbit coupling parameters are included in Table 13. Table 14 compares the experimental and theoretically calculated values[305] of F^i and λ for U^{3+}. There are large discrepancies, especially in F^2, which is about half the calculated value. Table 11 contains the data concerning the crystal-field parameters on $U^{3+}:LaCl_3$ and $Pu^{3+}:LaCl_3$[314] derived from the same experiments.

CHAPTER 3

Survey of Theory

3.1. Localized Electron Theories

In this section we review theoretical models of actinide compounds that are based on the assumption of $5f$ electrons well localized on the actinide ions. In these theories the influence of the ions (and, where present, the conduction electrons) surrounding the actinide ion is treated as an electrostatic potential which perturbs the intra-atomic actinide potential. This is the *crystal-field model*, which has been very successful in explaining the magnetic properties of compounds of $3d$ transition metals and of the $4f$ rare earths.

3.1.1. Assumptions of the Crystal-Field Model

Let us begin with a free actinide atom. There is a strongly bound Radon core ([Rn]) and several external $5f$, $6d$, and $7s$ electrons. For example, in the case of uranium the configuration is $[Rn]5f^36d^17s^2$. When a collection of these atoms and atoms of other elements are brought together to form a compound, the comparatively weakly bound $6d$ and $7s$ electrons, and perhaps also a $5f$ electron, tend to leave the parent actinide atom. They may be localized around the anion, as in the case of the ionic insulators (e.g., the halides), or they may form narrow $6d$ and broader $7s$ conduction bands, as in the case of the metallic compounds (see Fig. 93, top). Band electrons are also called itinerant electrons. Because of overlap with the wavefunctions of electrons centered on neighboring atoms, the discrete atomic energy levels are spread out into energy bands, the widths of which increase with decreasing atomic nearest-neighbor separation as shown in Fig. 93, top. The f levels spread out less rapidly than the $6d$ and $7s$ levels because of the smaller radial extension of the f wavefunctions.

In the case of most rare earth metals and compounds, the $4f$ levels remain

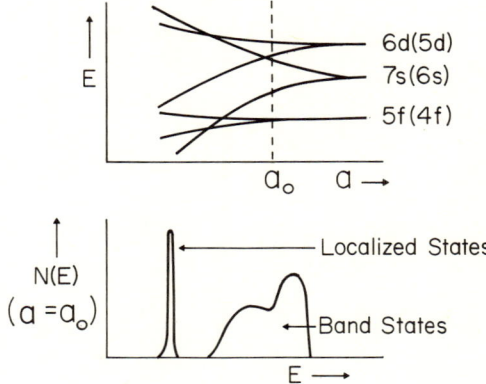

Fig. 93. Hypothetical model of a metallic actinide (rare earth) material for which the crystal-field treatment of the $5f$ ($4f$) electrons would be valid. *Top*: Position and splitting into bands of the levels as a function of the interatomic separation a. The equilibrium lattice constant is a_0. *Bottom*: Density of states $N(E)$ as function of electronic energy E.

very atomiclike at the equilibrium lattice constant a_0 and lie well below the bottom of the band as shown in Fig. 93, bottom. This is why the crystal-field theory with its assumptions of localized atomiclike f states weakly perturbed by the crystal field works so well for the rare earths. As we shall see later in the section on band theories (Sec. 3.2), however, the $5f$ wavefunctions in the lighter actinide metals and some intermetallics overlap so strongly with those of the $5f$, $6d$, and $7s$ electrons on neighboring atoms that the f states are fully itinerant, in which case the crystal-field description is inapplicable.

The effects of broadening of the $5f$ levels and their proximity to the $6d$–$7s$ band states will be thoroughly explored in a later section. For the present, we assume that the picture of Fig. 93, bottom, is valid for actinide compounds. Note that the $5f$ energies plotted in this figure do not have the meaning of single-particle energies (as do band energies), because the $5f$ localized states are highly correlated. The energy difference between a $5f$ localized state and a band state represents the energy required to remove one electron from an actinide ion and place it in the band state in question. This energy will in general depend on the relative occupation of the band and localized states. For the crystal-field model to be valid, the gap between the f levels and the bottom of the band must be large compared to $k_B T$. Otherwise some f electrons may be

3.1. LOCALIZED ELECTRON THEORIES

thermally excited to band states, leaving behind actinide ions of a higher valency. Since the thermal excitation is a random process, the valency or electronic configuration (e.g., $5f^3$) of a given magnetic ion becomes undetermined, and the crystal-field model cannot be applied.

Formally, it is possible to describe a mixed-valence state by an ionic wave function which is the linear superposition, with coefficients α, β, of states with different configurations, e.g., $\psi = \alpha|5f^3\rangle + \beta|5f^26d^1\rangle$. However, it is doubtful that crystal-field theory can be applied to such a state. In fact, the very interaction which creates the mixed valency is excluded from consideration in crystal-field theory, since in the latter either one or another pure valency has minimal energy — and is thus realized — but not the mixture of two.

Returning now to the assumptions of the crystal-field model, one should realize that the level scheme of Fig. 93, bottom, refers only to the actinide ion and not to the compound as a whole. While in a pure elemental solid there are no other levels, in a compound one has to include the energy levels of the anions as well. This gives a band scheme, which derives from the bonding and antibonding orbitals of all ions in the unit cell. An example of such a proposed band scheme is shown in Fig. 94. This scheme is suggested on the basis of optical spectroscopy of UAs.[272] The $4p$ band derives from the As ion, the free atom configuration being $[Ar]3d^{10}4s^24p^3$. The $5f$ and $6d$ bands originate from the U ion. The $6d$ electrons, having a larger mean radius than the $5f$ electrons, are subject to a stronger crystal field than the latter, and their band is split in two subbands in the cubic field (e_g and t_{2g}). Other bands are not shown: the $7s$ band

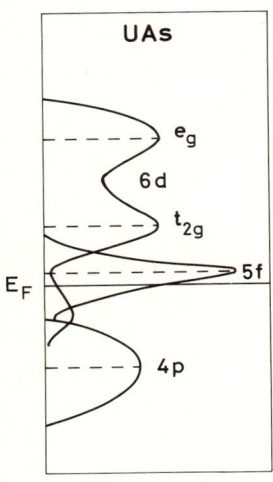

Fig. 94. Energy-level scheme of UAs as derived from optical spectroscopy. From Ref. 272. The vertical axis represents the energy, and in the horizontal direction to the right is plotted the density of states. The letters and numbers denote the band character and E_F = Fermi energy.

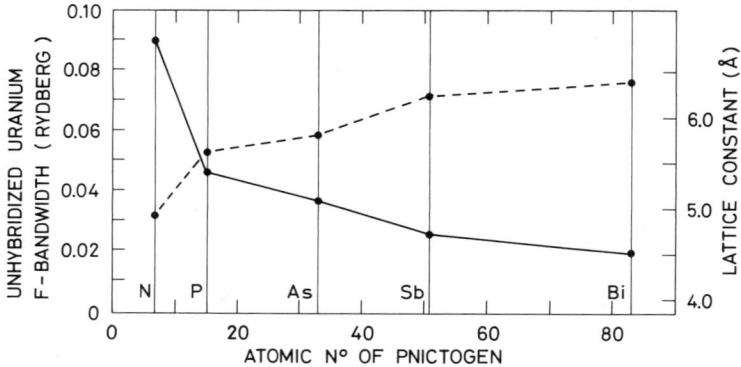

Fig. 95. Unhybridized uranium f-bandwidth and lattice constant (broken line) as a function of pnictogen atomic number. From Ref. 306.

of uranium is very broad measured on the scale of this figure, and extends presumably from slightly below E_F upwards.

Calculations confirm the clear correlation between lattice constant and $5f$ bandwidth implied in Fig. 93.[306] This can be seen in Fig. 95 where the f bandwidth is plotted as a function of pnictogen atomic number for the compounds UX (X = pnictogen). If one is interested in the magnetic properties only, which are due to the actinide $5f$ electrons, the knowledge of the complete band structure is not necessary.

The Hamiltonian of the $5f$ electrons in the crystal-field model is written as follows:

$$\mathcal{H}_{5f} = \mathcal{H}_0 + \mathcal{H}_{coul} + \mathcal{H}_{so} + \mathcal{H}_{cf} + \mathcal{H}_{mag}, \qquad (3.1.1\text{-}1)$$

where $\mathcal{H}_0$ is the kinetic energy and the potential energy in the field of the core electrons. The intra-atomic Coulomb interaction $\mathcal{H}_{coul}$ and the spin–orbit term $\mathcal{H}_{so}$ for the ion in question can sometimes be determined from spectroscopic measurements as described in Chapter 2. Because of the symmetry of the crystalline environment, the crystal-field term $\mathcal{H}_{cf}$ can be written in terms of a few parameters and spherical harmonic functions of the $5f$-electron coordinates giving the angular dependence of the effective crystalline electrostatic field experienced by the f electrons.

In applications to the rare earths, the $4f$ electrons, lying deep within the outer s, p, and d shells (their root mean square distance R from the nucleus being of order $0.5\,\text{Å}^{307}$), are well shielded from the crystal field. Thus, the latter field can be treated as a perturbation acting on the $4f$ electrons, which in the absence of the crystal field form an eigenstate of definite total angular momentum J of the ionic Hamiltonian (including the large spin–orbit term). The wavefunctions of the $5f$ electrons of actinides are much less well shielded ($R \simeq 0.7$ Å[308]) and experience a strong crystal field of the order of 1000 cm^{-1} (see Table 11). Thus, one must be prepared to diagonalize the Hamiltonian exactly with respect to a basis of the eigenstates of several J manifolds.[309]

The last term $\mathcal{H}_{mag}$ in the Hamiltonian of $5f$ electrons, Eq. (3.1.1-1), is due to interionic interactions. It is responsible for the type of magnetic order which may occur spontaneously below a certain critical temperature. Section 3.3 describes different forms of $\mathcal{H}_{mag}$.

3.1.2. Crystal-Field Theory

In this section we shall discuss what can be inferred from crystal-field theory about the properties of actinide compounds.

Crystal-field theory (CFT) commonly designates a method to determine the eigenfunctions and energy levels of a single ion in its crystalline environment, and the subsequent calculation of observables (such as, the magnetic susceptibility as a function of temperature) with the help of these functions and energies.

The fundamental assumption of CFT is that the electrons in an unfilled shell of the ion (which are responsible for most of the observed phenomena) may be treated as moving in an electrostatic potential created by all other charged particles in the crystal, except the ones localized on the same ion. The interaction with the latter, i.e., with the nucleus and other electrons of the same ion, is treated more completely, for instance, taking into account the Pauli principle and relativistic effects.

CFT neglects the correlation between electrons on different ions. Consequently, it will yield sharp and separated energy levels for the ion as long as the potential gives rise to bound states. Correlation between electrons on different ions gives rise to broad levels — the electronic energy bands of the crystal.

It follows, therefore, that CFT may be applied as a first approach to a given problem if the observed levels are sharp, whereas band theory is more appropriate if the levels are broad, that is, if the level breadths are comparable to level separations. In addition, as has been pointed out in Section 3.1.1., it is necessary that $k_B T$ be small compared to the energy separation between the f level and the bottom of a band of itinerant electrons so that the ion will be in a definite configuration f^n.

A penetrating discussion of the possible validity of CFT has been given in a series of papers by Freeman and Watson.[310,311] These authors discuss how the effect of the correlation between electrons on different ions may be considered as being due to covalency, i.e., the partial transfer of electrons from other ions to the one in question, and to overlap, i.e., the overlap of wavefunctions on different ions.

Many of these considerations are probably relevant to the actinides, but they cannot, at present, be quantitatively assessed. At many instances in this review, evidence is presented for the sharpness of the $5f$ levels of actinide ions, and this justifies the use of CFT as a tool for a first analysis of experimental results in these cases. A discussion of typical cases of success and of failure of CFT will conclude this section.

Once the fundamental assumption of CFT is accepted, several stages of its application may be envisaged.

In the most basic approach, the ionic energy levels would be calculated by solving the Dirac equation, or the Schrödinger equation with spin-orbit coupling for the many-electron wavefunctions of the ion in a specific crystal-field potential. To our knowledge, such calculations have not been done. (For the free ions, see, e.g., the early relativistic calculations of Lewis et al.[405].)

A simpler approach may be used, the nature of which, however, depends on the type of unfilled shell of the ion. In this approach, one considers the electrons in the unfilled shell as moving in the combined potential of the nucleus and core electrons, to which are added the spin-orbit potential and the crystal-field potential.

As pointed out in Section 3.1.1., in the transition metal series ($3d^n$) and rare earth series ($4f^n$) one *or* the other of the latter two potentials may be treated as a perturbation. In the actinides one is compelled to diagonalize the total Hamiltonian of the electrons in the unfilled $5f$ shell with respect to a manifold of states with different total angular momenta J. One immediate consequence of this is that the final states will be mixtures of states with different angular momenta, and the energy-level diagrams derived by Lea, Leask, and Wolf[312] for the f^n configuration in cubic crystal fields become inapplicable: the energies of the crystal-field levels will depend on the crystal-field parameters in a manner different from that shown in these diagrams. This has been first pointed out by Erdös and Rudra[309] and subsequently taken into account by Chan and Lam.[313]

The theory of crystal-field splittings of atomic-energy levels was given by Stevens,[314] and it has been summarized by Fick and Joos.[315] Besides the calculations of Lea et al.,[312] Chan and Lam have carried out calculations for f^n configurations for octahedral ("positive") crystal fields, and Erdös and

3.1. LOCALIZED ELECTRON THEORIES

Razafimandimby[316] for cubal ("negative") crystal-field configurations. Further developments of CFT were discussed by Judd.[317]

In the following section we will develop CFT as needed for the understanding of this review and present some explicit results as examples. We will also explain the role of shielding. We will then discuss the calculation by CFT of certain observables that are relevant to the magnetic properties of actinide compounds.

3.1.2a. Mathematical Formulation

The following facts are known from the theory of free many-electron atoms (see, e.g., Stevenson[376]).

All full shells create spherically symmetric electrostatic potentials, in which the electrons of the unfilled shells may be assumed to move.

The sum of relativistic (spin-dependent) interaction energies of an electron in an unfilled shell with those in a filled shell is zero.

These two facts about the free ions are taken over as assumptions regarding the ions in crystals. Whether these assumptions are justified will be discussed in the section about shielding.

We will also make use of the following assumption: The energy of interaction of the electrons in the unfilled $5f$ shell with the electrons in other unfilled shells is negligible. These other unfilled shells are, in the actinide series, $6d$ and $7s$ shells. This last assumption means that the $6d$ and $7s$ electrons have either been transferred to an anion, or are itinerant and do not hybridize with the $5f$ electrons. Thus, they may be considered as tributaries to the crystal field.

In formulating crystal-field theory for the ions of the iron and rare earth group elements, one usually starts with the following Hamiltonian

$$\mathcal{H} = \mathcal{H}_{\text{coul}} + \mathcal{H}_{\text{so}} + \mathcal{H}_{\text{cf}}. \qquad (3.1.2\text{-}1)$$

Here, $\mathcal{H}_{\text{coul}}$ represents the Coulomb interaction Hamiltonian of the electrons in the unfilled shell with the core and with each other. $\mathcal{H}_{\text{so}}$ is the spin-orbit interaction Hamiltonian, and $\mathcal{H}_{\text{cf}}$ is the crystal-field Hamiltonian. We shall discuss these three terms in the sequel.

It has been found that the low-lying levels of free ions can be satisfactorily reproduced by theory neglecting correlations and configurational interaction. That is to say, the following factors are taken into account:

a. The Coulomb interaction of the electrons in the open shells with a spherically symmetric core;

b. the Coulomb interaction of pairs of electrons in the open shells; and
c. the spin–orbit coupling of the electrons in the open shells.

The interaction (a) adds the same constant to all energy levels. The interactions (b) and (c) are dealt with in the following subsections.

3.1.2b. The Coloumb Hamiltonian

The Coulomb interaction of electrons in the open shells is taken into account in the following way: The neglect of correlation implies that the eigenfunctions of $\mathcal{H}$ will be linear combinations of Slater determinants which, in turn, are sums of products of single-electron wavefunctions. Therefore, the Coulomb interaction operator is simply a scalar (αr_{ij}^{-1}; r_{ij} = distance between electrons i and j) whose matrix element between two states ψ and ψ' will vanish if the total spins S and S' of the two states are different.

As shown, for instance, by Slater, the physical explanation for this is that the torques two electrons exert on each other are internal to the system and do not change the vector sum of their angular momenta. The matrix elements are also zero, if J differs from J'. Therefore $\mathcal{H}_{\text{coul}}$ is diagonal with respect to S, L, and J, i.e., its matrix in the $(\alpha LSJM)$ representation (see p. 110) is diagonal, except possibly with respect to the remaining quantum numbers, collectively denoted by α (e.g., seniority):

$$(l^n \alpha LSJM | \mathcal{H}_{\text{coul}} | l^n \alpha' L'S'J'M') = \delta_{LL'} \delta_{SS'} \delta_{JJ'} \delta_{MM'} f(\alpha, \alpha', L, S) \qquad (3.1.2\text{-}2)$$

These matrix elements can be, for the f^n configuration, expressed as numerical multiples of the Racah parameters E^i

$$f(\alpha, \alpha', L, S) = \sum_{i=0}^{4} a_i(\alpha, \alpha', L, S) E^i \qquad (3.1.2\text{-}3)$$

The factors a_i are tabulated by Nielson and Koster.[318] The Racah parameters themselves are certain linear combinations of the four Slater integrals F^i ($i = 1, 2, 4, 6$), defined as

$$F^i = 2 \int_0^\infty \int_0^\infty |u^0(5f; r_1)|^2 |u^0(5f; r_2)|^2 (r_<^k/r_>^{k+1}) r_1^2 r_2^2 dr_1 dr_2 \qquad (3.1.2\text{-}4)$$

and

$$E^0 = F^0 - 2F^2/45 - F^6/33 - 50F^6/1287,$$

$$E^1 = 14F^2/405 + 7F^6/297 + 350F^6/11583,$$

3.1. LOCALIZED ELECTRON THEORIES

$$E^2 = F^2/2025 - F^4/3267 + 175F^6/1656369,$$

$$E^3 = F^2/135 + 2F^4/1089 - 175F^6/42471.$$

$u^0(5f; r)$ is the radial factor of the wavefunction of the $5f$ electron, and $r_<$ is the smaller, $r_>$ the larger, of r_1 and r_2.

The experimental uncertainties in E^i do not have much consequence on the crystal-field levels originating from the lowest free-ion level, as long as this level arises mainly from a single multiplet, and as long as the magnitude of the crystal field is not comparable to the separation to the next multiplet.

The parameters F^i cannot, at present, be calculated satisfactorily theoretically, as can be seen on the example of U^{3+} from Table 12. Calculated and experimental values differ by as much as 45%. For this reason, these parameters are considered as parameters to be determined empirically from spectroscopic data (see Section 2.9.).

The form of Eqs. (3.1.2-2 and 3.2.2-3) of the Coulomb Hamiltonian is the one used in most crystal-field calculations where the energies and wave functions of the low levels are needed to calculate such observables as susceptibility, etc. For such purposes, the levels with energies between 0 and a few kT are sufficient.

To fit spectroscopic data, more elaborate Hamiltonians are used. Even though they include terms which are not strictly Coulombic in origin, we will list them here. Second-order perturbation theory yields several corrections which are additive to the energy. These are:

$$\mathcal{H}' = \alpha L(L+1) + \beta G(G_2) + \gamma G(R_7) + \Sigma T^i t_i + E_M. \quad (3.1.2\text{-}5)$$

The first three terms arise from interconfigurational interactions: When taking into account the Coulomb interaction between an electron in the shell l^n and an electron in another shell beyond the mean field approximation, second-order perturbation theory yields diagonal matrix elements which involve states corresponding to the promotion of an electron from the closed shell to the shell l. $G(G_2)$ and $G(R_7)$ are the eigenvalues of the Casimir operator for the groups R_7 and G_2 used to classify states of f-electron atoms. These eigenvalues are uniquely determined by the quantum numbers which characterize the state (just as for the cofactor of α), and their expression is given in Ref. 319. The next to last terms, where $i = 2, 3, 4, 6, 7,$ and 8, are terms which arise when the promotion of two electrons to shell l is taken into account. Finally, the last term E_M includes six further parameters, called M^k ($k = 0, 2, 6$) and p^k. The quantities M^k are Marvin integrals over the radial functions

$$M^k = \frac{e^2\hbar^2}{8m^2c^2}\langle(nl)^2| \frac{r_<^k}{r_>^{k+3}} |(nl)^2\rangle \qquad (3.1.2\text{-}6)$$

where $r_</r_>$ has the same meaning as in the Slater integrals, and P^k ($k = 0, 2, 4$) are coefficients of the so-called electrostatically correlated spin–orbit interaction.[320]

They enter into the expressions which describe the spin–spin and spin-other-orbit interactions, whose operators in this approximation can be expressed as

$$\mathcal{H}_{ss} = -\rho[(\mathbf{L}\cdot\mathbf{S})^2 + \tfrac{1}{2}(\mathbf{L}\cdot\mathbf{S}) - \tfrac{1}{3}L(L+1)S(S+1)] \qquad (3.1.2\text{-}7)$$

and

$$\mathcal{H}_{soo} = \lambda' \mathbf{L}\cdot\mathbf{S}. \qquad (3.1.2\text{-}8)$$

All these interactions seem sufficiently important to be included if one aims at reproducing the energy levels with an accuracy of $\cong 20\,\text{cm}^{-1}$. For instance, for Np^{3+}, $\alpha = 31.5\,\text{cm}^{-1}$, $\beta = -740\,\text{cm}^{-1}$, $\gamma = 899\,\text{cm}^{-1}$, $T^2 = 278\,\text{cm}^{-1}$, $M^0 = 0.68\,\text{cm}^{-1}$, and $P^2 = 896\,\text{cm}^{-1}$.[301]

3.1.2c. The Spin-Orbit Hamiltonian

$\mathcal{H}_{so} = \Sigma \gamma_i \mathbf{l}_i \cdot \mathbf{s}_i$ is the spin-orbit interaction energy, where $\mathbf{l}_i$ and $\mathbf{s}_i$ are the orbital and spin–angular-momentum operators of electron i, and the sum is over all electrons in the shell.

A few words about $\mathcal{H}_{so}$ are in order. For equivalent electrons, e.g., for the f^n configuration, the coefficients γ_i ($i = 1, \ldots, n$) are equal, and will be denoted by γ. However, it is only for a given multiplet, i.e., given L and S, that one can write $\mathcal{H}_{so} = \mathcal{H}_{LS}$ where $\mathcal{H}_{LS}$ is defined as

$$\mathcal{H}_{LS} = \lambda \mathbf{L}\cdot\mathbf{S}, \quad \mathbf{L} = \Sigma \mathbf{l}_i, \quad \mathbf{S} = \Sigma \mathbf{s}_i, \qquad (3.1.2\text{-}9)$$

since, in general, λ is different for each multiplet. Hence using this form of $\mathcal{H}_{so}$ with a simple coefficient λ will introduce an inaccuracy into the results if more than one multiplet has to be taken into account in the diagonalization of $\mathcal{H}$. Examination of the $5f^n$ configuration shows that the ground state contains a mixture of states which have L and S values other than those of the lowest multiplet. Specifically, in f^2, the ground state is 91% $^3H_4$ and 8.3% $^1G_4$ and 0.7% $^3F_4$, and the first excited state at $3346\,\text{cm}^{-1}$ consists of 82.7% $^3F_2$, 1.1% $^3P_2$, and 16.2% $^1D_2$. The admixtures to the ground state increase with the f shell being closer to being half-filled (reaching $\cong 50\%$ nonHund contribution for f^6) as noted by Chan and Lam.[313] Another example of this mixing, the ion U^{3+}, is presented in Table 15.

Table 15. Composition of the Eigenstates Z_i of the Lowest-Lying Manifold of Crystal-Field States of the U^{3+} Ion, as Observed[a] in a $LaCl_3$ Host. (The manifold arises mainly from the $^4I_{9/2}$ state of the free ion. Energies in cm^{-1}.)

Z_1 $0\ cm^{-1}$	Z_2 $207\ cm^{-1}$	Z_3 $245\ cm^{-1}$	Z_4 $427\ cm^{-1}$	Z_5 $451\ cm^{-1}$
$-0.863\ ^4I_{9/2}\ (-7/2)$	$-9.16\ ^4I_{9/2}\ (1/2)$	$-0.720\ ^4I_{9/2}\ (-9/2)$	$-0.850\ ^4I_{9/2}\ (5/2)$	$-0.716\ ^4I_{9/2}\ (3/2)$
$+0.331\ ^2H2_{9/2}\ (-7/2)$	$+0.362\ ^2H2_{9/2}\ (1/2)$	$-0.562\ ^4I_{9/2}\ (3/2)$	$+0.350\ ^2H2_{9/2}\ (5/2)$	$+0.561\ ^4I_{9/2}\ (-9/2)$
$-0.323\ ^4I_{9/2}\ (5/2)$	$-0.114\ ^2H1_{9/2}\ (1/2)$	$+0.286\ ^2H2_{9/2}\ (-9/2)$	$+0.315\ ^4I_{9/2}\ (-7/2)$	$+0.295\ ^2H2_{9/2}\ (3/2)$
$+0.124\ ^2H2_{9/2}\ (5/2)$		$+0.223\ ^2H2_{9/2}\ (3/2)$	$-0.130\ ^2H2_{9/2}\ (-7/2)$	$-0.231\ ^2H2_{9/2}\ (-9/2)$
$-0.105\ ^2H1_{9/2}\ (-7/2)$			$-0.112\ ^2H1_{9/2}\ (5/2)$	
98.5%	98.3%	96.6%	97.4%	96.8%

[a] From Ref. 303. The bracketed number is J^z. The contributions listed add up to the percentage shown.

To treat the spin–orbit interaction correctly, it is most convenient to express $\mathcal{H}_{so}$ in terms of double (i.e., orbital and spin) tensor operators

$$V_{p,q}^{1k} = \sum_i s_p^1(i) u_q^k(i), \quad p = -1, 0, 1; \quad q = -k, \ldots, k. \quad (3.1.2\text{-}10)$$

The meaning of these operators is explained in Appendix A. A complete set of states (not necessarily eigenstates of $\mathcal{H}$) will be needed to express all matrix elements of $\mathcal{H}$. As such it is customary to use the Russell–Saunders states

$$|l^n \alpha SLJM)$$

labelled by the following quantum numbers:

l = orbital angular momentum of the shell (for f electrons $l = 3$)
n = number of electrons in the shell
α = symbolic notation for all other quantum numbers necessary to uniquely label the state besides l, n, S, L, J, and M. None is needed for $n < 3$. The seniority quantum number is needed for $n \geq 3$, and others are needed for $n > 3$.
S = modulus of total spin **S**
L = modulus of total orbital momentum **L**
J = modulus of the total angular momentum $\mathbf{J} = \mathbf{L} + \mathbf{S}$
M = azimuthal quantum number, i.e., z component of **J**

With this labelling, the matrix elements of the spin orbit Hamiltonian are found to be (see Appendix A)

$$(l^n \alpha SLJM | \mathcal{H}_{so} | l^{n'} \alpha' S'L'J'M') = \gamma \delta_{nn'} \delta_{JJ'} \delta_{MM'} (-1)^{S'+L+J'}$$

$$\times \sqrt{l(l+1)(2l+1)} \begin{Bmatrix} J & L & S \\ 1 & S' & L' \end{Bmatrix} (l^n \alpha SL \| V^{11} \| l^n \alpha' S'L'). \quad (3.1.2\text{-}11)$$

Here the operator V^{11} is defined as

$$V^{11} = -V_{-1,1}^{11} + V_{0,0}^{11} - V_{1,-1}^{11}, \quad (3.1.2\text{-}12)$$

the curly bracket is a $6j$ symbol (Racah coefficient)[†], and $(l^n \alpha SL \| V^{11} \| l^n \alpha' S'L')$ is a reduced matrix element. The $6j$ symbols are most extensively tabulated in Landolt-Börnstein,[321] the reduced matrix elements in Nielson and Koster.[318]

Within a multiplet, i.e., for $\alpha' = \alpha$, $L' = L$, $S' = S$, Eq. (3.1.2-11) reduces to a numerical multiple of the matrix elements of the simplified spin–orbit operator $H_{LS} = \lambda \mathbf{L} \cdot \mathbf{S}$. This yields a relation between λ and γ of the form

---

[†] We use the definitions as given by A. R. Edmonds.[329] This treatise contains a comparative table of other frequently used definitions as well (e.g., Rose,[331] etc.).

3.1. LOCALIZED ELECTRON THEORIES

$$\lambda = 2(-1)^{S+L+J}\sqrt{l(l+1)(2l+1)}\begin{Bmatrix} J & L & S \\ 1 & S & L \end{Bmatrix}(l^n\alpha SL\|V^{11}\|l^n\alpha SL)$$

$$\times [J(J+1)-L(L+1)-S(S+1)]^{-1}\gamma. \qquad (3.1.2\text{-}13)$$

However, the latter operator incorrectly gives zero matrix elements between different multiplets.

The Hamiltonian of a free ion (ion in vacuum) consists of $\mathcal{H}_{coul} + \mathcal{H}_{so}$. Hence the Slater parameters F_i together with the spin–orbit coupling constant γ can be obtained by matching the spectroscopically obtained energy levels of the free ion with those calculated by diagonalizing $\mathcal{H}_{free}$ using different trial sets of parameters. In practice, the free ion situation is approximated by placing the ion in a crystal producing little crystal field and being transparent (e.g., CaF_2) or in aqueous solution.

3.1.2d. The Crystal-Field Hamiltonian

The electrons in the unfilled shell of a given ion find themselves in the electrostatic potential produced by all other charges in the crystal. Excluding the nucleus and closed shells of the free ion in question – whose potential is included in $\mathcal{H}_{coul}$ – these charges, with density $\rho(x')$, give rise to a potential energy V given by

$$V = e\sum_{i=1}^{n}\int\Psi_i^*(x_i)\frac{\rho(x')}{|x_i-x'|}d^3x'\,\Psi_i(x_i)d^3x_i \qquad (3.1.2\text{-}14)$$

since the charge density of the electrons in the unfilled shell is $e\Sigma_k\Psi^*\Psi$. Hence, V may be regarded as the average of the operator

$$\mathcal{H}_{cf}(x_1,\ldots,x_n) = e\sum_{i=1}^{n}\int\frac{\rho(x')}{|x_i-x'|}d^3x'. \qquad (3.1.2\text{-}15)$$

We introduce the spherical polar coordinates r_i, θ_i, ϕ_i of the electron i at x_i with reference to a coordinate system centered at the ion whose Hamiltonian is $\mathcal{H}_{cf}$. As is well known, the Coulomb potential (Eq. 3.1.2-14) can be expanded as follows:

$$\mathcal{H}_{cf} = \sum_{k=1}^{\infty}\sum_{m=-k}^{k}\sum_{i=1}^{n}B_k^m r_i^k Y_k^m(\theta_i,\phi_i). \qquad (3.1.2\text{-}16)$$

The functions Y_k^m are normalized spherical harmonics. The superscript k denotes the kth power of r.

The quantities B_k^m, which represent different moments of the charge distribution $\rho(\mathbf{x})$, are the crystal-field parameters. Their meaning may be elucidated as follows.

If the ion at $r = 0$ (but without its ith electron) could be removed from the crystal without modifying the distribution of other charges, the electron i would have the potential energy given by Eq. (3.1.2-16) (omitting the sum over i) where the coefficients B_k^m are replaced by another set of constants A_k^m. The latter are called *bare crystal-field coefficients*. If the ion is now thought to be replaced into the crystal, the total interaction energy of the electron i with the charges becomes that given by Eq. (3.1.2-16) (omitting the sum over i) involving the *shielded crystal-field coefficients* B_k^m.

Since $\mathcal{H}_{\text{coul}} + \mathcal{H}_{\text{so}}$ contains the interaction energy of the electron i with the nucleus and all other electrons of the ion at $r = 0$ in the absence of the crystal field, the change $A_k^m \to B_k^m$ represents the modification of the crystal field produced by the other ions through the polarization of the charge distribution of the ion to which the electron i belongs.

It is customary to set

$$B_k^m = A_k^m(1 - \sigma_k) \qquad (3.1.2\text{-}17)$$

and to call the σ_k (which turn out to be independent of m) the *Sternheimer shielding factors*.[322] (For U^{4+} and U^{3+} they have been calculated by Erdös and Kang,[323] and are given in Table 16. See also Section 3.1.2i – "Shielding.")

Table 16. Sternheimer Shielding Factors and Their Principal Direct (D) and Exchange (E) Contributions for Uranium Ions[a]

	U^{4+}	U^{3+}
σ_{2D} ($6s \to d$)	0.21	0.30
σ_{2D} ($6p \to f$)	0.92	0.74
σ_{2E} (all)	−0.27	−0.20
σ_2 (*total*)	*0.89*	*0.83*
σ_{4D} ($6p \to f$)	0.325	0.244
σ_{4D} ($6p \to h$)	0.060	0.056
σ_{4E} (all)	−0.321	−0.320
σ_4 (*total*)	*0.012*	*0.026*
σ_{6D} ($6p \to h$)	0.025	0.020
σ_{6D} ($6p \to j$)	0.013	0.011
σ_{6E} (all)	−0.094	−0.077
σ_6 (*total*)	*−0.046*	*−0.039*

[a] From Ref. 323.

3.1. LOCALIZED ELECTRON THEORIES

In general, it has not been possible to calculate reliably B_k^m for actinide compounds. They are to be regarded as empirical constants to be obtained from experiment (as explained further along). However, they are subject to certain restrictions arising from (i) the requirements that the potential obey Laplace's equation, and (ii) the type of point symmetry which may exist around $r = 0$. Requirement (i) is automatically fulfilled using Eq. (3.1.2-16) in which $B_k^m Y_k^m$ is multiplied not by an arbitrary function of r, but by r^k.

3.1.2e. Point Symmetry

Let us abbreviate (θ_i, ϕ_i) by i. Since $\mathcal{H}_{cf}$ is a scalar it has to be invariant with respect to all operations of the point symmetry group G of the ion. It is well known that under rotations and rotary reflections the spherical harmonics of a given k transform into linear combinations of each other. Hence, for each k, only those linear combinations of the $Y_k^m(i)$ will occur in $\mathcal{H}_{cf}$, which are invariant under all operations of G. These are the linear combinations which transform according to a unit (i.e., one-dimensional) irreducible representation $D^\mu(k)$ of G. These unit representations are obtained by decomposing the $(2k + 1)$ dimensional representation $D(k)$ of G in the space spanned by the linearly independent functions Y_k^m ($m = -k, \ldots, k$). The index μ numbers these unit representations: $\mu = 1, \ldots, s(k)$. Their number $s(k)$ is the multiplicity of the unit representation.

The above-mentioned linear combinations, $s(k)$ in number, are known and given for each k and for each point-group G in Table 17 (see p. 114).

Let us denote these invariant linear combinations by

$$Y_{k\mu}(G, i) = \sum_{m=-k}^{k} C_{k\mu}^m(G) Y_k^m(i) \quad (3.1.2\text{-}18)$$

where the index G reminds us that for each group the coefficients are different. It can be seen from Table 17 that these coefficients are 0, 1, or -1, except in the cubic groups.

We note that for the cubic groups, the functions $Y_{k\mu}$ are the Kubic harmonics.[324] From what has been said above, it follows that $\mathcal{H}_{cf}$ may now be written as

$$\mathcal{H}_{cf} = \sum_{i=1}^{n} \sum_{k=1}^{\infty} \sum_{\mu=1}^{s(k)} r_i^k B_{k\mu}(G) Y_{k\mu}(G; i). \quad (3.1.2\text{-}19)$$

Since most frequently $s(k)$ is much less than $2k + 1$, for a given k the number of the newly introduced symmetry-adapted crystal-field coefficients $B_{k\mu}(G)$ is usually smaller than the number $(2k + 1)$ of B_k^m.

Table 17. General Form of the Crystal-Field Potential for Electrons with $l \leq 3$.[a] (For explanation see section "Point group symmetry.")

Point group	Crystal-Field potential
O, O_h, T_d	$B_4(Y_4^0 + \sqrt{\frac{5}{14}}Y_4^4 + \sqrt{\frac{5}{14}}Y_4^{-4}), B_6(Y_6^0 - \sqrt{\frac{7}{2}}Y_6^4 - \sqrt{\frac{7}{2}}Y_6^{-4})$
T, T_h	$B_4(Y_4^0 + \sqrt{\frac{5}{14}}Y_4^4 + \sqrt{\frac{5}{14}}Y_4^{-4}), B_{61}(Y_6^0 - \sqrt{\frac{7}{2}}Y_6^4 - \sqrt{\frac{7}{2}}Y_6^{-4}) + B_{62}(Y_6^2 + Y_6^{-2} - \sqrt{\frac{5}{11}}Y_6^6 - \sqrt{\frac{5}{11}}Y_6^{-6})$
$C_{6v}, D_6, D_{6h}, D_{3h}$	$B_2Y_2^0, B_4Y_4^0, B_{61}Y_6^0 + B_{62}(Y_6^6 + Y_6^{-6})$
C_6, C_{6h}, D_{3h}	$B_2Y_2^0, B_4Y_4^0, B_{61}Y_6^0 + B_{62}Y_6^6 + B_{62}^*Y_6^{-6}$
$C_{4v}, D_4, D_{4h}, D_{2d}$	$B_2Y_2^0, B_{41}Y_4^0 + B_{42}(Y_4^4 + Y_4^{-4}), B_{61}Y_6^0 + B_{62}(Y_6^4 + Y_6^{-4})$
C_4, C_{4h}, S_4	$B_2Y_2^0, B_{41}Y_4^0 + B_{42}Y_4^4 + B_{42}^*Y_4^{-4}, B_{61}Y_6^0 + B_{62}Y_6^4 + B_{62}^*Y_6^{-4}$
C_{3v}	$B_2Y_2^0 + B_{41}Y_4^0 + B_{42}(Y_4^3 - Y_4^{-3}), B_{61}Y_6^0 + B_{62}(Y_6^3 - Y_6^{-3}) + B_{63}(Y_6^6 + Y_6^{-6})$
D_3	$B_2Y_2^0, B_{41}Y_4^0 + iB_{42}(Y_4^3 + Y_4^{-3}), B_{61}Y_6^0 + iB_{62}(Y_6^3 + Y_6^{-3}) + B_{63}(Y_6^6 + Y_6^{-6})$
C_3	$B_2Y_2^0, B_{41}Y_4^0 + B_{42}Y_4^3 - B_{42}^*Y_4^{-3}, B_{61}Y_6^0 + B_{62}Y_6^3 - B_{62}^*Y_6^{-3} + B_{63}Y_6^6 + B_{63}^*Y_6^{-6}$
C_{2v}, D_2, D_{2h}	$B_{21}Y_2^0 + B_{22}(Y_2^2 + Y_2^{-2}), B_{41}Y_4^0 + B_{42}(Y_4^2 + Y_4^{-2}) + B_{43}(Y_4^4 + Y_4^{-4}), B_{61}Y_6^0 + B_{62}(Y_6^2 + Y_6^{-2}) + B_{63}(Y_6^4 + Y_6^{-4}) + B_{64}(Y_6^6 + Y_6^{-6})$
C_{1h}, C_2, C_{2h}	$B_{21}Y_2^0 + B_{22}Y_2^2 - B_{22}^*Y_2^{-2}, B_{41}Y_4^0 + B_{42}Y_4^2 + B_{42}^*Y_4^{-2} + B_{43}Y_4^4 + B_{43}^*Y_4^{-4}, B_{61}Y_6^0 + B_{62}Y_6^2 + B_{62}^*Y_6^{-2} + B_{63}Y_6^4 + B_{63}^*Y_6^{-4} + B_{64}Y_6^6 + B_{64}^*Y_6^{-6}$
C, C_i	All coefficients present; (Coeff. of Y_k^{-m}) = $(-1)^m$(Coeff. of Y_k^m)

[a] Table compiled by P. Monachesi.

3.1. LOCALIZED ELECTRON THEORIES

Comparing Eqs. (3.1.2-16) and (3.1.2-19) and using (3.1.2-18) it follows that the two kinds of crystal-field coefficients are related as follows

$$B_k^m = \sum_{\mu=1}^{s(k)} B_{k\mu}(G) C_{k\mu}^m. \quad (3.1.2\text{-}20)$$

Consider an example: In the case of O_h-symmetry, group theory yields $s(l \text{ odd}) = 0$, $s(2) = 0$, $s(4) = 1$. Hence $B_1^m = B_2^m = B_3^m = 0$, and $B_4^m = B_4(O_h) C_4^m$. Consequently, the ratio $B_4^m/B_4^{m'} = C_4^m/C_4^{m'}$ is completely determined by the symmetry alone. For example, $B_4^4/B_4^0 = \sqrt{5/14}$ for $G = O_h$. It may be possible (and examples exist in the literature) to fit certain experimental observations by disregarding this relation and using the ratio B_4^4/B_4^0 as an adjustable parameter. This, however, is tantamount to abandoning the crystal-field concept.

We give in Table 17 the linear combinations of spherical harmonics which appear in Eqs. (3.1.2- 20 and 25) as coefficients of r_i^k (or $\langle r^k \rangle$) for all crystallographic point groups.[†] The following rules were adopted for the table: a) The sign convention for the spherical harmonics is that of Ballhausen;[304] b) The second subscript of $B_{k\mu}$ is left off if there is only one coefficient, i.e., if $s(k) = 1$; and c) All $B_{k\mu}$ are real unless the complex conjugate $B_{k\mu}^*$ appears in the same expression.

3.1.2f. Matrix Elements of $\mathcal{H}_{cf}$

The wavefunction of the n electrons in the unfilled (here $5f$) shell, subject only to the Coulomb field of the core, may be written as

$$\Psi(r_1\theta_1\phi_1, \ldots, r_n\theta_n\phi_n) = u^0(5f; r_1) \ldots u^0(5f; r_n)|l^n\alpha LSJM\rangle. \quad (3.1.2\text{-}21)$$

Here, $u^0(5f; r_k)$ is the radial factor of the one-electron $5f$ wavefunction, and $|l^n\alpha LSJM\rangle$ is the angular and spin factor of Ψ, as defined in the section on the spin–orbit Hamiltonian. Using Eq. (3.1.2-19), the matrix element of $\mathcal{H}_{cf}$ between two such wavefunctions Ψ and Ψ' is

$$(\Psi'|\mathcal{H}_{cf}|\Psi) = \sum_k \sum_\mu \sum_i B_{k\mu}(G)\langle r^k \rangle (l^n\alpha L'S'J'M'|Y_{k\mu}(G;i)|l^n\alpha LSJM),$$
$$(3.1.2\text{-}22)$$

where

$$\langle r^k \rangle = \int_0^\infty r^k |u^0(5f; r)|^2 dr \quad (3.1.2\text{-}23)$$

[†] We are grateful to Dr. P. Monachesi for the group-theoretical determination of these expressions. See also Ref. 396.

is an average of the kth power of the distance of the $5f$ electron from the nucleus. It is now possible to introduce a *new crystal-field Hamiltonian operator* $\tilde{\mathcal{H}}_{cf}$, *from which the radial coordinates have been eliminated*, defined by

$$\tilde{\mathcal{H}}_{cf} = \sum_k \sum_\mu B_{k\mu}(G) \langle r^k \rangle \sum_i Y_{k\mu}(G;i). \qquad (3.1.2\text{-}24)$$

This operator acts only on the angular factors $|l^n \alpha LSJM\rangle$ of the wavefunctions. It is seen from Eq. (3.1.2-22) that the crystal-field coefficients occur always in the combination $B_{k\mu}(G)\langle r^k \rangle$. Therefore only the value of these products determines the physical properties and can be deduced from experiment. The radial integrals $\langle r^k \rangle$ cannot be calculated accurately for ions in a crystal since, especially for high k, the wavefunctions may be greatly expanded or contracted with respect to their rather precisely known shape in a free ion.

It proved to be most expeditious to express the spherical harmonics in Eq. (3.1.2-24) by Racah's unit tensor operators, the matrix elements of which have been tabulated.[318] Using the definition, Eq. (A-4) of the n-electron unit tensor operator $\mathbf{U}_m^k$, as well as Eqs. (3.1.2-18) and (A-9), we arrive at the final result

$$(l^n \alpha LSJM |\tilde{\mathcal{H}}_{cf}| l^n \alpha' L'S'J'M') = \delta_{SS'}(-1)^{S+L'+2J-M+l}(2l+1)$$

$$\times \sqrt{(2J+1)(2J'+1)/4\pi} \sum_k \begin{pmatrix} l & l & k \\ 0 & 0 & 0 \end{pmatrix} \begin{pmatrix} J & k & J' \\ -M & M-M' & M' \end{pmatrix} \begin{Bmatrix} L & J & S \\ J' & L' & k \end{Bmatrix}$$

$$\times \langle r^k \rangle (l^n \alpha SL \| \mathbf{U}^k \| l^n \alpha' SL') \sum_\mu C_{k\mu}^{M'-M} B_{k\mu}(G). \qquad (3.1.2\text{-}25)$$

We note that the summands over k are nonzero only if k is even, $k \leq J+J'$, $k \leq 2l$ and $|M'-M| \leq k$. We also note that by virtue of Eq. (3.1.2-20) the sum over μ equals $B_k^{M'-M}$. This eliminates all point symmetry relations from the matrix elements, but one has to bear in mind that the coefficients $B_k^{M'-M}$ are not linearly independent.

We recall that the two-story round brackets are the Wigner $3j$ symbols (or Clebsch–Gordan coefficients), and the curly bracket is the $6j$ symbol or Racah coefficient.[321] (See footnote on p. 110.)

As an example for the application of the methods of the foregoing and of this paragraph, we present the results of calculations concerning the splitting of the $^3H_4$ level of U^{4+} in an octahedral and a cubal crystal field, respectively (Fig. 96). The two differ in the relative sign of the coefficients V_6 and V_4, which are the sixth- and fourth-order crystal-field coefficients (cf., Table 17, first row). These splittings have already been calculated for the octahedral field ($V_6/V_4 > 0$).[313]

3.1. LOCALIZED ELECTRON THEORIES

The slight differences between the work of S. K. Chan and D. J. Lam[313] and that of P. Erdös and H. A. Razafimandimby[316] are due to the fact that the latter authors used a new set of Slater parameters,[325] which are in agreement with the experimental data concerning the magnetic susceptibility of UCl_4.[326]

3.1.2g. The Zeeman Hamiltonian

The effect of an external magnetic field **H** can easily be incorporated into the calculation of the crystal-field levels by including in the Hamiltonian the term

$$\mathcal{H}_z = \boldsymbol{\mu} \cdot \mathbf{H}, \quad \boldsymbol{\mu} = g\mu_B(\mathbf{L} + 2\mathbf{S}). \quad (3.1.2\text{-}26)$$

If the external field can be taken to coincide with the quantization axis (z axis), the matrix of $\mathcal{H}_z$ (i.e., of μ_z) expressed with respect to the states $|f^n \alpha SLJM\rangle$ is diagonal with regard to f^n, α, L, S, and M. Abbreviating the state by $|J\rangle$, one finds

$$(J|\mu_z|J) = gM, \quad g = 1 + [J(J+1) - L(L+1) - S(S+1)]/2J(J+1)$$

$$(3.1.2\text{-}27)$$

and

$$(J|\mu_z|J-1) = (J-1|\mu_z|J)$$

$$= \frac{1}{2}\left[\frac{(J^2 - M^2)(S+L+J+1)(S+L-J+1)(S+J-L)(L+J+S)}{J^2(2J+1)(2J-1)}\right]^{1/2}. \quad (3.1.2\text{-}28)$$

If H_x and H_y do not vanish, the matrix elements of μ_x and μ_y have to be taken into account as well. These formulas may be obtained by expressing the vector $\boldsymbol{\mu}$ as a spherical tensor of rank 1, and applying the method outlined in the Appendix. The final results are summarized by S. K. Chan and D. J. Lam.[313]

3.1.2h. Thermodynamic Averages

Once the energy levels of the ion in the crystal field are known, the free energy F can be calculated from:

$$F = -k_B T \ln \sum_j \exp(-\beta E_j), \quad \beta = (k_B T)^{-1}, \quad (3.1.2\text{-}29)$$

where the sum extends over all levels known. From the free energy all other thermodynamic functions may be derived, as for instance, the magnetization **M** and the susceptibility χ:

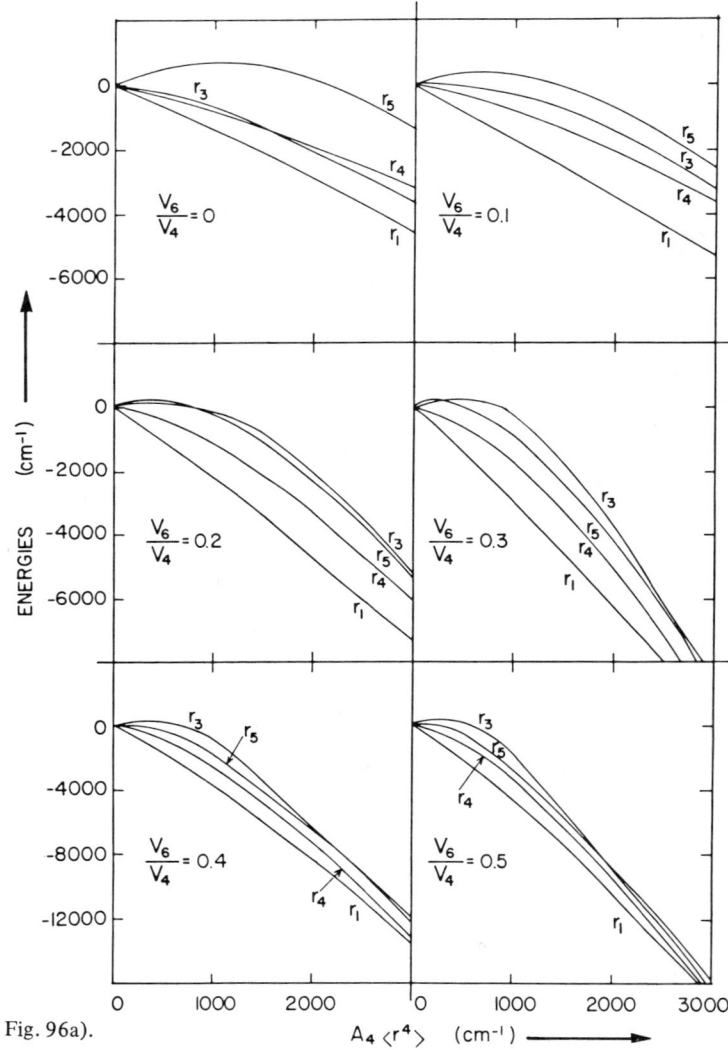

Fig. 96. Energy-level diagram of the lowest multiplet $^3H_4$ of the ion U^{4+} (f^2 configuration) in a crystal field of cubic symmetry. a). octahedral surroundings, $V_6/V_4 > 0$. b). cubal surroundings, $V_6/V_4 < 0$. V_6 and V_4 are the crystal field coefficients of sixth and fourth order, respectively (cf. Table 17). The symbols Γ_i denote the irreducible representations of the cubic group. The Coulomb- and spin-orbit parameters used are from Ref. 325. From Ref. 316.

3.1. LOCALIZED ELECTRON THEORIES

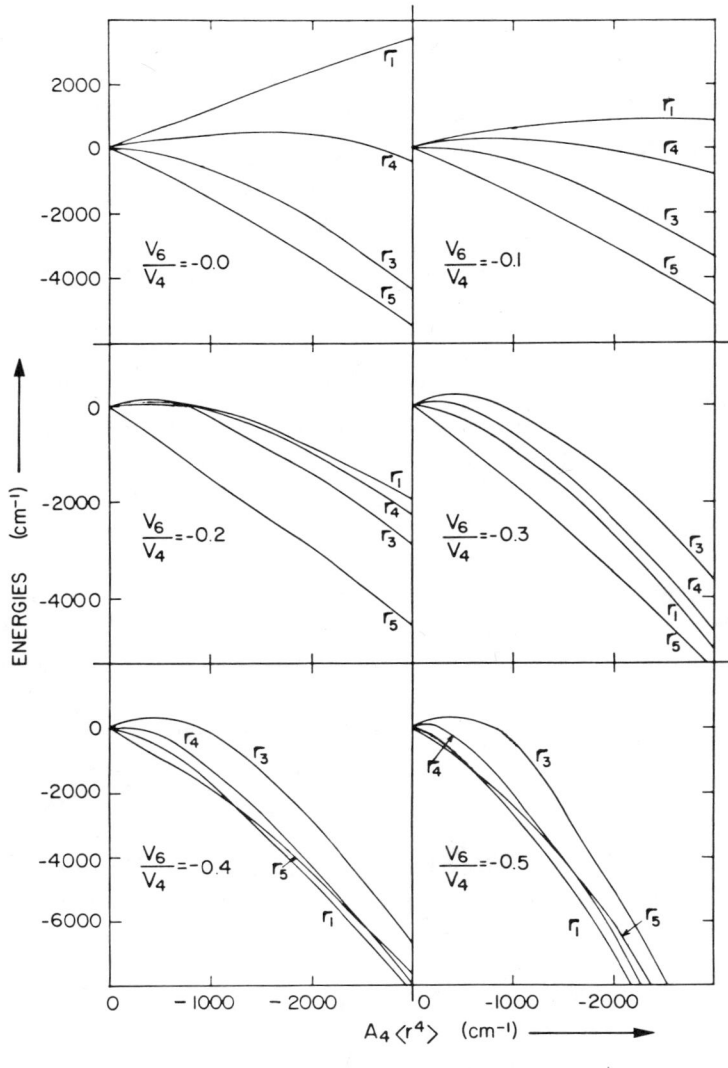

Fig. 96b).

$$\mathbf{M} = -\partial F/\partial \mathbf{H}, \quad \chi_{ij} = \frac{\partial M_i}{\partial H_j}, \quad i,j = x, y, z. \qquad (3.1.2\text{-}30)$$

A more direct method to calculate the magnetization is to evaluate the sum

$$\mathbf{M}(T) = \sum (\Psi_i|\mu_B(\mathbf{L} + 2\mathbf{S})|\Psi_i) \exp(-\beta E_i). \qquad (3.1.2\text{-}31)$$

Since in the process of diagonalization of $\mathcal{H}$ the state vectors Ψ_i are calculated as linear combinations of $|l^n\alpha LSJM)$-states, $\mathbf{M}$ can be calculated in the same computational process using the coefficients $a(l^n\alpha LSJM)$ of these linear combinations, i.e.,

$$\Psi_i = \sum_{\alpha JSJM} a(l^n\alpha LSJM)|l^n\alpha LSJM) \qquad (3.1.2\text{-}32)$$

and

$$\mathbf{M} = \sum a^*(l^n\alpha LSJM)a(l^n\alpha'L'S'J'M')(l^n\alpha LSJM|\mu_B(\mathbf{L} + 2\mathbf{S})|l^n\alpha'L'S'J'M').$$
$$(3.1.2\text{-}33)$$

The matrix elements in the last expression are the same as the ones introduced in the section "Zeeman Hamiltonian."

This allows the calculation of $\mathbf{M}$ (and of χ) as a function of the adjustable parameters, such as the crystal-field coefficients.

3.1.2i. Shielding

As expounded in the section on crystal-field calculations, the bare crystal field, i.e., the field produced by the distribution of charges other than the ones belonging to the ion in question, is modified at the site of the f electron due to shielding. This effect was first investigated in detail by Sternheimer for certain ions.[322] The accuracy of different calculational methods was compared by Erdös and Kang.[327] The shielding factors σ_i ($i = 2, 4$, and 6) for the ions U^{3+} and U^{4+} have been calculated by Erdös and Kang.[323] For other actinide ions they have not been calculated. The shielding factors σ^i arise as the sum of contributions due to different excitations, which Sternheimer classified as *direct* $\sigma_D(nl_i \overset{k}{\to} l_f)$ and *exchange* $\sigma_E(nl_i \overset{k}{\to} l_f; l_o)$.[327] In Table 16 we give the largest contributions, as well as the sum total of σ_D, σ_E, and finally $\sigma = \sigma_D + \sigma_E$. It is seen that for U^{4+}, $\sigma_2 = 0.888$, $\sigma_4 = 0.012$, and $\sigma_6 = 0.046$.

In cubic (cubal, octa-, and tetrahedral) point symmetry, the second-order crystal field vanishes, hence σ_2 does not enter into play, and σ_4 and σ_6, which are a few percent, have a physical meaning.[†]

The factor σ_2, which becomes important for noncubic symmetry, is of the order 1, and its actual value is rather uncertain for the following reasons:

[†] This may be asserted even though large cancellations occur in the sum (see Table 16), because the individual terms were calculated to a 6-digit accuracy.

3.1. LOCALIZED ELECTRON THEORIES

a. The shielding factors are calculated in first-order perturbation theory. The results will be acceptable if the bare crystal-field potential is much smaller than the Coulomb potential due to the nucleus and closed shells;

b. The calculations use the unperturbed wavefunctions (in this case the Hartree–Fock–Slater wavefunctions worked out by Lenander).[328] As can be expected and as is seen from Table 16, the largest contributions to the shielding factors come from the open $6s$ and $6p$ shells exterior to the $5f$ shell in the *free ion*. These contributions may be afflicted with considerable errors in the *crystal*.

Therefore, if the crystal field is strong, the shielding affects the crystal-field coefficient B_2^m to an unknown degree, and may even change its sign. This shielding effect, along with covalency and overlap, excludes meaningful point-charge calculations of crystal fields in actinide compounds (see, e.g., Ref. 410).

3.1.2j. Comments on the Crystal-Field Theory

It remains an open question, whether the most important parameters F^i ($i = 0, 2, 4, 6$) and the spin–orbit coupling constant γ, derived for the free ion by fitting (e.g., over a hundred spectroscopic levels up to 20,000 cm^{-1}), are suitable for use in calculations concerning low-lying thermally-excitable crystal-field levels. It would seem that parameters obtained by fitting fewer and lower-lying levels, neglecting configurational and other corrections of the type included in $\mathcal{H}'$, which introduce many additional parameters (see Eq. (3.1.2-5)), would be more suitable.

We note at this point that a change in the parametrization of the crystal field for the actinides has been proposed by Judd.[330] This author ascribes different radial function $u^0(5f; r_i)$ to each $5f$ orbital. It follows that in Eq. (3.1.2-24) one has to replace $\langle r^k \rangle$ by $\langle r_i^k \rangle$, and the subsequent equations have to be modified, giving rise to more independent crystal-field coefficients than indicated here. This procedure has not yet been adequately tested.

Another remark concerns the Stevens operator equivalent method.[314] In this method, in the crystal-field Hamiltonian

$$\widetilde{\mathcal{H}}_{cf} = \sum B_l^m \langle r^k \rangle Y_l^m, \qquad (3.1.2\text{-}34)$$

the spherical harmonics are replaced by certain linear combinations, O_l^m of products of the components J_i ($i = x, y, z$ or $+, -, z$) of the total angular momentum operator:

$$\tilde{\mathcal{H}}_{cf} = \sum B_l'^m \langle r^k \rangle O_l^m. \qquad (3.1.2\text{-}35)$$

The definition of these operators O_l^m may be found, for instance, in Ref. 312. Their advantage lies in the fact that the matrix elements of angular-momentum operators are easily found.

Despite their usefulness for rare earths, for actinides the use of these operators is not recommended. This is due to the fact that in contrast to the B_l^m, the coefficients $B_l'^m$ (W and x for octahedral fields)[312] depend on the quantum numbers LSJ. Therefore, as soon as the free ion states are mixtures of states with different J, the m coefficients $B_l'^m$ lose their meaning. Even if one uses the $|LSJ\rangle$ basis, for each J a different set of $B_l'^m$ coefficients would have to be introduced. An example of this mixture is given in Table 15.

Two examples of the application of crystal-field theory (CFT) are described in Section 2.9, Spectroscopic Data. In one example, 45% of the 250 calculated levels could be assigned to the levels deduced from the absorption and fluorescence spectra of Np^{3+}:$LaCl_3$. On the other hand, in the reproduction of experimental results concerning magnetic properties, CFT achieved only partial success.

For instance, the monopnictide UN has been theoretically analyzed[332] by assuming a cubic crystal-field potential containing terms up to $k = 4$ (see Table 17). The neglect of the term $k = 6$ seems appropriate, since in the other pnictides this term is small: In USb neutron data are best fitted with $V_4 = -300$ K, $V_6 = -15$ K,[333] and in UP and UAs the values $V_4 = 3200$ K and $V_6 = 30$ K have been suggested[334] ($V_k = \langle r^k \rangle B_k$). A further assumption of the above-mentioned analysis was that the f^2 ground state is a pure $J = 4$ level split by the crystal field. This assumption is supported by the fact that best fit to data in UX-type compounds has been achieved for USb with pure-J models, whereas the assumption of J-mixing worsened the fit.[333]

With this pure-J model the relative magnetization vs. temperature is calculated, but yields a curve which is up to 15% above the experimental one. The calculated paramagnetic susceptibility vs. temperature has the wrong curvature below 200 K, and using first and second nearest-neighbors Heisenberg exchange, the resulting sign of the second-nearest-neighbor exchange constant is inconsistent with the observed AFM-I-type ordering.

Recent neutron scattering investigations[335] on UAs show that the basic interactions in monopnictides are more complex than assumed in simple CFT.

3.1. LOCALIZED ELECTRON THEORIES

For instance, the exchange interaction is highly anisotropic, being $\cong 40$ times stronger between the (001) planes than in the planes. The observed oblate shape of the quadrupole moment of the $5f$ electrons — if it also exists in UN — excludes the type of wavefunction assumed in the simple CFT analysis.[334]

From this example, one may conclude that CFT is an important ingredient of theories of many actinide compounds, but has to be refined and supplemented by detailed microscopic models of interactions.

In the so-called mixed-valence compounds where the energies of two configurations of f electrons are equal, one would not expect CFT to be applicable. It is remarkable, however, that even in these cases the application of CFT concepts to the two participating configurations as incorporated in a more general theory leads to results in good agreement with experiment. This has been shown, for instance, for UP[16] and for the rare earth compound TmSe.[336]

3.1.3. Interionic Interactions

In the following models a, b, c, and d of interionic interactions we write the interaction Hamiltonian between the localized ions as

$$\mathcal{H} = \sum_{i,j} \mathcal{H}_{ij},$$

where the summation is over all pairs of sites i and j of the crystal occupied by actinide ions.

3.1.3a. Heisenberg Exchange

We recall that for the insulating compounds of $3d$ metals, the orbital contribution to the magnetic moment of the $3d$ electrons is usually "quenched" by the strong crystal fields of about $10{,}000\,\mathrm{cm}^{-1}$. The resulting exchange interaction between a pair of magnetic ions at lattice sites i and j is then well described by the isotropic Heisenberg exchange Hamiltonian $\mathcal{H}^{\mathrm{ex}}$ given by

$$\mathcal{H}^{\mathrm{ex}}_{ij} = J_{ij} \mathbf{S}_i \cdot \mathbf{S}_j . \tag{3.1.3-1}$$

Here, $\mathbf{S}_i$ is the spin operator of ion i, and the coupling constant J_{ij} falls off exponentially and may also vary in sign with the distance between the sites. Eq. (3.1.3-1) also holds for metal ions interacting via intervening anions, in which case one speaks of the Anderson superexchange.[337] The magnitudes of J_{ij} are hard to estimate for actinide compounds because of the presence of several other kinds of exchange interaction (see below), but values of about

$k_B \times 7$ K for nearest neighbor coupling of U^{4+} ions in UO_2 have been deduced from fitting inelastic neutron scattering[25] and far infrared absorption data.[14]

3.1.3b. Anisotropic Exchange

In the compounds of $4f$ and $5f$ metals, the large spin–orbit coupling indicates that the orbital contribution to the magnetic moment is only partially reduced by the crystal field. This means, because of the highly directional nature of the f orbitals, that the exchange between two f ions may be expected to contain anisotropic terms — that is, to depend on the angles between the magnetic moments and the crystallographic axes as well as on the relative angle between the two moment vectors. The presence of anisotropic interactions and electrostatic multipole interactions between pairs of rare-earth ions has been shown experimentally by Birgeneau and others.[338] Quadratic anisotropic exchange between a pair i, j of ions takes the form

$$\mathcal{H}_{ij}^{an} = \sum_{\mu\nu} K_{ij}^{\mu\nu} J_i^\mu J_j^\nu, \tag{3.1.3-2}$$

where J_i^μ ($\mu = x, y,$ or z) is the μ component of the angular momentum operator $\mathbf{J}_i$ of the actinide ion at site i and $K_{ij}^{\mu\nu}$ is the anisotropic exchange constant.

For actinide ions, few estimates for these anisotropic exchange constants are available. For illustration, in Table 18 we give these interaction constants derived from measurements[338] for a rare earth ion. A schematic illustration of the consequences of anisotropic exchange is shown in Fig. 97.

Table 18. Two Possible Theoretical Solutions for the Anisotropic Interaction Constants in cm^{-1} for Ce^{3+} Pairs in $LaCl_3$ and in $LaBr_3$.[a] $H_{12} = \mathbf{S}_1 \cdot \mathbf{K} \cdot \mathbf{S}_2$. (The magnetic dipole–dipole (MDD) contributions are also listed. The estimated errors in the calculated values are ± 0.03. $K_\perp = \frac{1}{2}(K_{xx} + K_{yy})$. The superscripts indicate nearest neighbor (nn) and next nearest neighbor (nnn).)

	Ce^{3+} in $LaCl_3$			Ce^{3+} in $LaBr_3$		
	MDD	#1	#2	MDD	#1	#2
K_{zz}^{nn}	−0.168	0.225	−0.015	−0.156	0.29	0.06
$K_\perp^{nn}$	0.000	0.015	−0.225	0.000	−0.03	−0.38
K_{zz}^{nnn}	0.024	−0.075	0.005	0.022	−0.08	0.00
$K_\perp^{nnn}$	0.000	−0.075	0.005	0.000	−0.08	0.00

[a] From Ref. 338.

3.1. LOCALIZED ELECTRON THEORIES

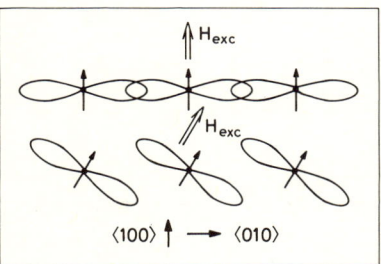

Fig. 97. Schematic illustration of anisotropic exchange in actinide compounds. If only this exchange were present, the actinide magnetic moments (arrows) would preferentially line up in the ⟨100⟩ direction, because the 5f electronic charge clouds shown overlap more and thus the exchange field H_{ex} is greater than if the moments point in, say, the ⟨110⟩ direction.

3.1.3c. Electric Multipole Interactions

The electrostatic quadrupole interactions contribute the term

$$\mathcal{H}_{ij}^{em} = \sum_{\mu\nu\rho\delta} A_{ij}^{\mu\nu\rho\delta} O_{i\mu\nu} O_{j\rho\delta}, \qquad (3.1.3\text{-}3)$$

where $O_{i\mu\nu}$ is the operator for the $\mu\nu$ component of the quadrupole moment of the ion at site i, and μ, ν, ρ, δ, each take on the values x, y, and z in the summation. The quantities $A_{ij}^{\mu\nu\rho\delta}$ are coupling constants between the ions at sites i and j. The multipole interaction of Eq. (3.1.3-3) plays a dominant role in Allen's[14] theory of UO_2 and in the Long–Wang theory[13] of UP and is discussed in more detail below.

3.1.3d. Biquadratic Exchange

The biquadratic exchange is normally written in the form

$$\mathcal{H}_{ij}^{bq} = K_{ij}(\mathbf{J}_i \cdot \mathbf{J}_j)^2, \qquad (3.1.3\text{-}4)$$

and has several possible origins. It occurs as a higher order perturbation correction to the Heisenberg exchange [Eq. (3.1.3-1)]. The quadrupole interaction [Eq. (3.1.3-3)] contains terms biquadratic in the operators $\mathbf{J}_i$ and $\mathbf{J}_j$. Hence Eq. (3.1.3-4) may be considered as a special case of Eq. (3.1.3-3). Also, some theories of *magnetoelastic coupling* lead to effective biquadratic exchange interaction.[339,340]

3.1.3e. RKKY Exchange

The localized magnetic moments of the actinide ions in metallic compounds may also be magnetically coupled indirectly through an exchange interaction $\mathcal{J}_{sf}$ with the itinerant band electrons, as occurs in the rare earth metals.

The interaction Hamiltonian $\mathcal{H}_{sf}$ of the n band electrons and N localized moments on sites $\mathbf{R}_i$ is given by

$$\mathcal{H}_{sf} = \sum_{\alpha=1}^{n} \sum_{n=1}^{N} \mathcal{J}_{sf}(\mathbf{r}_\alpha - \mathbf{R}_i)\mathbf{J}_i \cdot \mathbf{s}_\alpha, \qquad (3.1.3\text{-}5)$$

where $\mathbf{s}_\alpha$ is the Pauli vector spin operator of the electron α at position $\mathbf{r}_\alpha$. Eq. (3.1.3-5) can be rewritten in terms of a spatially varying effective magnetic field $\mathbf{h}(\mathbf{r})$ seen by the itinerant electrons. The field $\mathbf{h}(\mathbf{r})$ is given by the expression

$$\mathbf{h}(\mathbf{r}) = -\mu_0^{-1} \sum_i \mathcal{J}_{sf}(\mathbf{r} - \mathbf{R}_i)\mathbf{J}_i. \qquad (3.1.3\text{-}6)$$

In the Ruderman-Kittel-Kasuya-Yosida (RKKY) theory,[341] the interaction $\mathcal{J}_{sf}(\mathbf{r} - \mathbf{R}_i)$ is assumed to be proportional to a delta function of $(\mathbf{r} - \mathbf{R}_i)$. The delta-function form for $\mathcal{J}_{sf}(\mathbf{r} - \mathbf{R}_i)$ is more appropriate for the $4f$ electrons of the rare earths than for $5f$ electrons, because the wavefunctions of the $4f$ electrons are less spatially extended than those of the $5f$ electrons. Theories improving upon the RKKY assumption and investigating the resulting effects upon magnetic and electrical properties have appeared.[342-344]

The RKKY treatment of $\mathcal{H}_{sf}$ leads[341] to an effective isotropic Heisenberg-type exchange $\mathcal{H}_{ij}$ between two localized ions at sites i and j in a cubic lattice of lattice constant a_0 of the form:

$$\mathcal{H}_{ij} = \Gamma(2k_F a_0)^4 \frac{(x_{ij}\cos x_{ij} - \sin x_{ij})}{x_{ij}^4} \mathbf{J}_i \cdot \mathbf{J}_j, \qquad (3.1.3\text{-}7)$$

where Γ is a coupling constant, k_F is the Fermi wave number of the band electrons, and $x_{ij} = 2k_F|\mathbf{R}_i - \mathbf{R}_j|$. Both $\mathcal{J}_{sf}$ and $k_B T$ must be small compared to the Fermi level of the band for these results to be valid. Thus, if the number of band electrons and the Fermi level do not change over the temperature range of interest, the quantity multiplying $\mathbf{J}_i \cdot \mathbf{J}_j$ in Eq. (3.1.3-7) is a temperature-independent constant; and the s-f exchange can be absorbed in the Heisenberg exchange, Eq. (3.1.3-1). The RKKY coupling, however, has a very long range compared to the direct and superexchange interactions, which fall off exponentially with distance.

The RKKY interaction varies in sign as a function of x_{ij} (see Fig. 98). This allows a change in magnetic structure when the density of conduction electrons, and thereby k_F, changes. As can be seen in Fig. 98, a change in the spatial period of the conduction electron polarization may cause the neighboring ion spin to be in a region of electron-spin polarization opposite to the one before the change.

3.1. LOCALIZED ELECTRON THEORIES

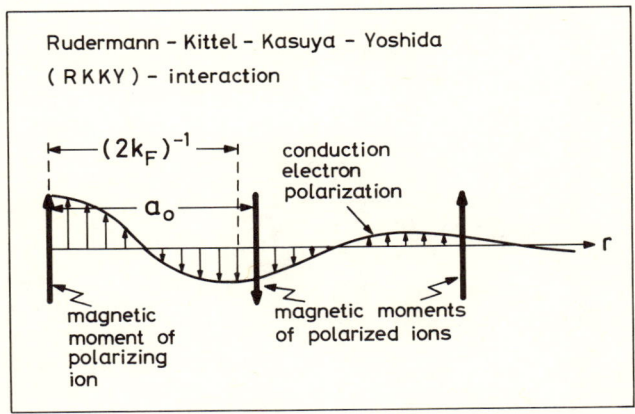

Fig. 98. Schematic diagram of the interaction of ionic magnetic moments via conduction electron spin polarization ($RKKY$ interaction). k_F = Fermi wave number, proportional to the 1/3 power of the conduction electron density; a_0 = distance between magnetic ions.

By replacing the operators $\mathbf{J}_i$ and $\mathbf{J}_j$ in Eq. (3.1.3-7) by their average values and summing over i and j, the classical magnetic ground-state energy F_R can be calculated for any type of magnetic ordering. The results[19] are shown in Fig. 99 as functions of the parameter $2k_F a_0$ for the case of the NaCl-type metallic actinide compounds. The most stable magnetic structure for a given value of $2k_F a_0$ has the lowest magnetic energy F_R. Assuming a parabolic band, we find

$$2k_F a_0 = 2(12\pi^2)^{1/3} w^{1/3} \quad (3.1.3\text{-}8)$$

in the NaCl lattice, where w is the number of band electrons per actinide ion. Thus, both the magnetic structure and the exchange coupling are expected to depend strongly on the band occupation in the metallic actinide compounds.

With reference to Fig. 99, the occurrence of ferromagnetism (FM) in the NaCl-type metallics US, USe, UTe, and NpC (below 200 K) may be explained[19] by assuming that $w \cong 2$ for these materials; whereas the presence of AFM-I ordering in UP, UN, and USb indicates $w \cong 1$. One should recall, however, that the probable presence of competing direct exchange or Anderson superexchange renders dubious the quantitative reliability of Fig. 99. Nevertheless, the changes in the type of magnetic ordering and in the Néel or Curie temperatures as the concentration x is varied in solid solutions such as $UP_{1-x}S_x$ can be qualitatively understood[16, 96, 97] from the RKKY calculation. With the assumption that the

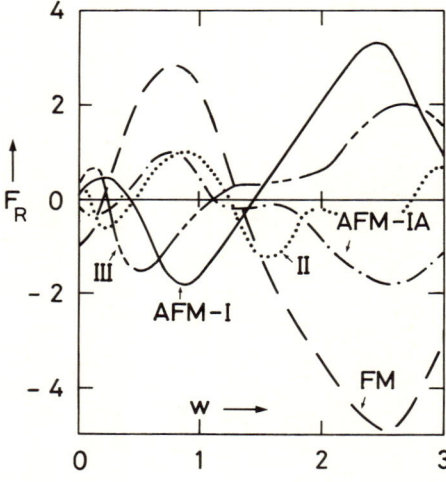

Fig. 99. The *RKKY* exchange energy F_R (in arbitrary units) as a function of the number w of band electrons per metal ion in the NaCl lattice, for various types of magnetic order. Refs. 16, 19.

band occupation w changes continuously from $w \cong 1$ to $w \cong 2$ as x varies from $x = 0$ to $x = 1$ in $UP_{1-x}S_x$, $UAs_{1-x}S_x$, $UAs_{1-x}Se_x$, and $UP_{1-x}Se_x$ (Figs. 40–47), the RKKY results predict that the exchange interaction F_R should first decrease with x until $x \cong 0.2$, at which point there should occur a change from antiferromagnetism to ferromagnetism followed by an increase in the exchange as x increases from $x \cong 0.2$ to $x \cong 1.0$. Since the ordering temperature is expected to be proportional to the exchange, these predictions correlate well with the observed decrease of T_N with x in the AFM regime and with the subsequent increase of T_C in the FM phase in Figs. 40–47. The RKKY calculation does not predict the occurrence of the AFM-IA structure observed in these phase diagrams and in UAs, because this phase never has the lowest exchange energy F_R although $F_R(w)$ has a shallow minimum near $w = 1.3$ (dash-dotted curve of Fig. 99). Kuznietz has proposed[96] that when the finite mean free path of the conduction electrons is considered, the AFM-IA phase can be the most stable phase at $w \cong 1.2$ between the large regions of stability of the AFM-I and FM phases. This picture is supported by the data of Figs. 40–47. A more detailed estimate of the number of conduction electrons at $T = 0$ K in UP, US, and UAs based upon an analysis of the phase diagrams is carried out in Ref. 16.

Adachi and Imoto[115] have carried out RKKY calculations for the cubic U_3X_4 and tetragonal UX_2 compounds discussed in Sections 2.1 and 2.2. Their results for the dependence of T_C (or θ) and T_N on $2k_F a$ are shown in Fig. 100 and account for the observed occurrence of the (+ – – +) and (+ – + –) orderings

3.1. LOCALIZED ELECTRON THEORIES

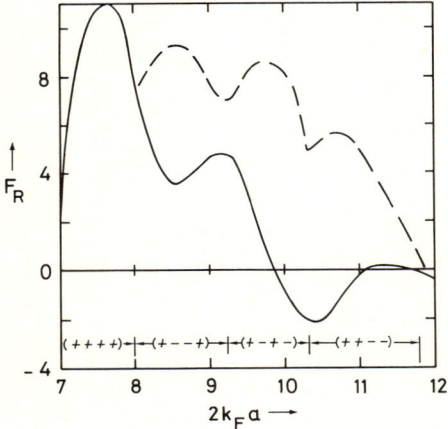

Fig. 100. The *RKKY* exchange energy F_R (in arbitrary units) as a function of $2k_F a$ in the cubic U_3X_4-type lattice (full curve) and the b.c. tetragonal UX_2-type lattice (dashed curve). k_F: Fermi wave number; a = lattice constant. The type of magnetic order and the corresponding region of $2k_F a$-values is indicated along the abscissa. Ref. 115.

listed in Table 4 for the UX_2 compounds. The tetragonal lattice of these compounds requires, of course, two lattice constants a and c for its specification, and the authors of Ref. 115 do not consider the dependence of their results on c, apparently making the approximation c/a = constant. This is true to within 2% for UP_2, UAs_2, UBi_2, and USb_2, but lattice summations can be sensitive to such changes. Adachi and Imoto also assume that the orbital magnetic moment is completely quenched in all metallic actinide compounds, which in view of the large spin-orbit coupling seems very dubious. Finally, these authors explain the variations of T_N and θ in the UX_2 and U_3X_4 compounds by making the assumption that the Fermi wavenumber k_F is constant for each class of compounds, so that $2k_F a$ depends only on a.

3.1.3f. Coqblin–Schrieffer Exchange

Coqblin and Schrieffer[345] (CS) showed that a weak hybridization between localized f states and itinerant band states also gives rise to an effective f-f exchange interaction via the scattering of itinerant electrons by the localized electrons. Even though Coqblin and Schrieffer worked out their theory for the rare earth ion Cerium rather than for an actinide ion, we will present here the essential features of this theory in view of its potential importance for the actinides.

The authors consider the simplest case, namely, a Ce^{3+} impurity ion in a metallic host. The ground state of Ce^{3+} is $4f^1$ ($j = 5/2$, $l = 3$), which is assumed to have energy E_0 relative to the Fermi level E_F of the conduction band. The Hamiltonian $\mathcal{H}$ of the system is

where
$$\mathcal{H} = \mathcal{H}_0 + \mathcal{H}_1, \quad (3.1.3\text{-}9)$$

and
$$\mathcal{H}_0 = \sum_{kM} \epsilon_k n_{kM} + E_0 \sum_M n_M + \tfrac{1}{2} U \sum_{MM'} n_M n_{M'}, \quad (3.1.3\text{-}10)$$

$$\mathcal{H}_1 = \sum_{kM} (V_k c^+_{kM} c_M + V_k^* c^+_M c_{kM}.) \quad (3.1.3\text{-}11)$$

The operators c_M^+ (c_M) create (destroy) a $4f$ electron having magnetic quantum number M ($M = -j, -j+1, \ldots j$) on the Ce^{3+} ion. The operators c_{kM}^+ (c_{kM}) create (destroy) a conduction electron in a $j = 5/2$ partial wave state of wavenumber k and angular momentum z-component $M\hbar$. The partial wave states are centered on the Ce^{3+} site and are linear combinations of plane wave states of wavevector **k** and spin σ; i.e.,

$$c_{kM}^+ = \sum_{\mathbf{k}\sigma} c_{\mathbf{k}\sigma} \langle kM | \mathbf{k}\sigma \rangle. \quad (3.1.3\text{-}12)$$

The energy required to localize two f electrons on the Ce ion is denoted by U; V_k is the mixing interaction between localized and band states, and n_{kM} (n_M) are number operators for electrons in the itinerant (localized) states.

The most important effect of $\mathcal{H}_1$ is exchange scattering, in which a conduction electron in the state $|kM\rangle$ is scattered into the state $|k'M'\rangle$ and simultaneously an f electron makes a transition from the localized state $|M'\rangle$ to the localized state $|M\rangle$. This process may be considered as proceeding by either one of two intermediate virtual states. First, an f electron in the state $|M'\rangle$ may be raised to the band, producing an extra band electron in the state $|k'M'\rangle$ and a hole at the Ce site (intermediate state no. 1). The mixing interaction then removes the electron in the band state $|kM\rangle$ and places it in the localized state $|M\rangle$. In the second possible process, the band electron in the state $|kM\rangle$ is removed and localized on the Ce ion, producing a state with two f electrons occupying the states $|M\rangle$ and $|M'\rangle$ and a hole in the band (intermediate state no. 2). The mixing interaction then transfers the electron in the state $|M'\rangle$ to the band state $|k'M'\rangle$. The authors now make a canonical transformation of $\mathcal{H}_1$ to a new interaction Hamiltonian $\mathcal{H}_2$ which is restricted to operate within the basis of four types of states: the initial and final states and the two intermediate states described above. One finds

$$\mathcal{H}_2 = -\sum_{kk'} \sum_{MM'} J_{kk'} c^+_{k'M'} c^+_M c_{M'} c_{kM}, \quad (3.1.3\text{-}13)$$

where $J_{kk'}$ is approximated as follows:

3.1. LOCALIZED ELECTRON THEORIES

$$J_{kk'} = \begin{cases} 0 & \text{if } |\epsilon_k| \text{ or } |\epsilon_{k'}| > D, \\ J & \text{otherwise.} \end{cases} \quad (3.1.3\text{-}14)$$

The quantity D is a cutoff parameter roughly equal to $|E_0|$, and

$$J = |V_{k_F}|^2 U/(E_0(E_0 + U)). \quad (3.1.3\text{-}15)$$

Transforming back to the conventional plane wave representation for the conduction band states, Coqblin and Schrieffer find

$$\mathcal{H}_2 = -\sum_{kk'}\sum_{\sigma\sigma'}\sum_{MM'} J^{MM'}_{k\sigma k'\sigma'} c^+_{k'\sigma} c_{k\sigma} \left(c^+_M c_{M'} - \frac{\delta_{MM'}}{2j+1}\sum_{M''} n_{M''} \right), \quad (3.1.3\text{-}16)$$

where the second term in the parentheses subtracts out the unimportant direct scattering $(M = M')$, and

$$J^{MM'}_{k\sigma k'\sigma'} = J_{kk'}\langle k\sigma|kM\rangle\langle k'M'|k'\sigma'\rangle. \quad (3.1.3\text{-}17)$$

The matrix elements in Eq. (3.1.3-17) may be obtained in terms of spherical harmonic functions of the angles of $\mathbf{k}$ and $\mathbf{k}'$ relative to the chosen reference system and in terms of angular momentum coupling coefficients dependent on j, M, and M'. The interaction $\mathcal{H}_{ij}$ between ions at sites $\mathbf{R}_i$ and $\mathbf{R}_j$ is as follows:

$$\mathcal{H}_{ij} = \sum_{k\sigma}\sum_{k'\sigma'}\langle k\sigma|\tilde{\mathcal{H}}_{ij}|k'\sigma'\rangle\langle k'\sigma'|\tilde{\mathcal{H}}_{ij}|k\sigma\rangle(\epsilon_k - \epsilon_{k'})^{-1}, \quad (3.1.3\text{-}18)$$

where

$$\tilde{\mathcal{H}}_{ij} = -\sum_{kk'}\sum_{\sigma\sigma'}\sum_{MM'}\sum_{n=i}^{j} J^{MM'}_{k\sigma k'\sigma'} e^{i(\mathbf{k}-\mathbf{k}')\cdot \mathbf{R}_n}\left(c^+_M(n)c_{M'}(n) - \frac{\delta_{MM'}}{2j+1}\sum_{M''} n_{M''}(n)\right)$$
$$\times c^+_{k'\sigma} c_{k\sigma}. \quad (3.1.3\text{-}19)$$

The interaction $\mathcal{H}_{ij}$ essentially results from considering all ways in which conduction electrons mediate an exchange of magnetic quantum numbers M and M' between the two impurity sites. Evaluating Eq. (3.1.3-19), and setting $\mathbf{R}_i - \mathbf{R}_j = \mathbf{R}_{ij}$, the authors arrive at

$$\mathcal{H}_{ij} = \sum_{MM'} E^{MM'}_{ij}(\mathbf{R}_{ij})\left(c^+_{M'}(i)c_M(i) - \frac{\delta_{MM'}}{2j+1}\sum_{M''} n_{M''}(i)\right)$$
$$\times \left(c^+_M(j)c_{M'}(j) - \frac{\delta_{MM'}}{2j+1}\sum_{M''} n_{M''}(j)\right), \quad (3.1.3\text{-}20)$$

where

$$E^{MM'}_{ij}(\mathbf{R}_{ij}) = 2\sum_{k\sigma}\sum_{k'\sigma'}\frac{f_k(1-f_{k'})}{\epsilon_k - \epsilon_{k'}}|J^{MM'}_{k\sigma k'\sigma'}|^2 \cos[(\mathbf{k}-\mathbf{k}')\cdot \mathbf{R}_{ij}]. \quad (3.1.3\text{-}21)$$

To the lowest order in $(k_F R_{ij})^{-1}$, the energy $E_{ij}^{MM'}$ has the following form:

$$E_{ij}^{MM'}(R_{ij}) = F(R_{ij})G(M,M'), \qquad (3.1.3\text{-}22)$$

where

$$F(R_{ij}) = m^* k_F^4 J^2 \pi^{-3} (2k_F R_{ij})^{-3} \cos(2k_F R_{ij}), \qquad (3.1.3\text{-}23)$$

with m^* the electron effective mass and $G(M,M')$ a function of M and M' only.

Siemann and Cooper[346,439] show that in a reference frame with the z axis along the line joining the Ce ions, $G(M,M') = 0$ unless $M = \pm 1/2$ and $M' = \pm 1/2$. Since $M = m_l + m_s$, we see that $m_l = m_l' = 0$ for a finite interaction. The latter authors provide the physical explanation that the strongest coupling is between f orbitals which pile up charge along the bonding axis, and these orbitals have $m_l = 0$ relative to that axis. Thus, the maximum orbital contribution to the ionic magnetic moment (and typically, therefore, the magnetic moment itself) will be perpendicular to the bonding axis. Thus, the anisotropic exchange mechanism acting between Ce^{3+} ions in the NaCl-type lattice energetically favors magnetic ordering along the [100] direction (see Fig. 97).

Assuming that these results also apply to the more complicated case of actinide ions, Cooper and Siemann can explain quantitatively the neutron scattering data[265,271] showing strong correlations within the ferromagnetic (001) planes perpendicular to the ⟨100⟩ magnetic axis of some actinide monopnictide antiferromagnets. The latter authors explain the ⟨111⟩ magnetic alignment observed in the ferromagnetic uranium chalcogenides as resulting from the assumed dominance of the crystal field, which favors the ⟨111⟩ orientation. To sum up, there is postulated a competition between the CS exchange, which favors ⟨100⟩ ordering, and the crystalline anisotropy, which favors ⟨111⟩. In the pnictides, the CS exchange is assumed to be the stronger influence, whereas the crystal field is stronger in the chalcogenides.

In an attempt to explain the recent observations of 2k and 3k magnetic structures in some uranium pnictides,[28,103] Cooper et al.[440–442] generalize $\mathcal{H}_{ij}$ (Eq. 3.1.3-20) to the case of an f^n ground-state configuration, with either an f^{n+1} or an f^{n-1} virtual intermediate state, and obtain

$$\mathcal{H}_{ij} = -F(\mathbf{R}_{ij}) \sum_{mm'} \left[\sum_{\alpha\alpha'} A_{mm'}^{\alpha\alpha'} \left(L_{\alpha'\alpha}^{(i)} - \frac{\delta_{\alpha\alpha'}}{2j+1} I^{(i)} \right) \right]$$

$$\times \left[\sum_{\beta\beta'} A_{mm'}^{\beta\beta'} \left(L_{\beta'\beta}^{(j)} - \frac{\delta_{\beta\beta'}}{2j+1} I^{(j)} \right) \right]. \qquad (3.1.3\text{-}24)$$

The operator $L_{\alpha'\alpha}^{(i)}$ is a standard basis[374] operator (see also Section 3.3) causing a transition of the ion at site i from state $|\alpha\rangle$ to state $|\alpha'\rangle$, and $I^{(i)}$ is the identity operator at site i. The coefficients $A_{mm'}^{\alpha\alpha'}$ are analogous to the functions $G(M,M')$ of Eq. (3.1.3-22). Guided by the discussion following the latter equation, Cooper et al. consider only the case $m = m' = \pm 1/2$ for the quantum number m_j of the transferred electron. In order to calculate the energy of various possible magnetic structures, it is necessary to transform the Hamiltonian $\mathcal{H}_{ij}$ to the crystalline coordinate system of the f.c.c. lattice. With the addition of a small Heisenberg exchange interaction between neighboring U ions, Thayamballi and Cooper[441] find that 2k and 3k magnetic structures indeed arise in their model for appropriate values of the parameters. It remains to be seen whether this approach can explain the details of the magnetization and susceptibility of UP and UAs as functions of temperature. In order to do this, it would seem to be necessary to incorporate the effects of the crystal field, on the ground state f^n configuration, because the crystal-field parameter V_4 is typically found to be $\cong 10^3$ K in the UX compounds.[334] An alternative theoretical approach based on anisotropic exchange and incorporating the crystal-field has been worked out by Monachesi and Weling,[452] and yields the 2k and 3k structures.

3.1.4. Theories of First-Order Transitions Based on Localized Models

3.1.4a. Biquadratic Exchange

As an illustration of how the presence of biquadratic exchange can lead to first-order transitions, consider a single $(2J + 1)$-fold degenerate ground state for each magnetic ion. In the "effective (or "molecular") field" approximation, each ion is assumed to be in an effective magnetic field h' proportional to the thermal average magnetization σ, self-consistently determined by the equation

$$\sigma = \frac{N}{V} \frac{\sum_{m=-J}^{J} \mu_0 m \exp(\mu_0 m h'/k_B T)}{\sum_{m=-J}^{J} \exp(-\mu_0 m h'/k_B T)} \quad (3.1.4\text{-}1)$$

where $\mu_0 = \mu_B g_J$, and g_J is the Landé factor. If the interaction is of the bilinear Heisenberg type [Eq. (3.1.3-1)], the effective field h' is simply $\lambda\sigma$, where λ is proportional to a sum of exchange parameters J_{ij}. In this case, there are at most two solutions for $\sigma(T)$ (Fig. 101, top left), and the solution of lowest free energy yields the well known Brillouin curve, with $\sigma(T)$ falling continuously to zero at T_N or T_C (Fig. 101, bottom left). In the presence of biquadratic exchange the effective field becomes

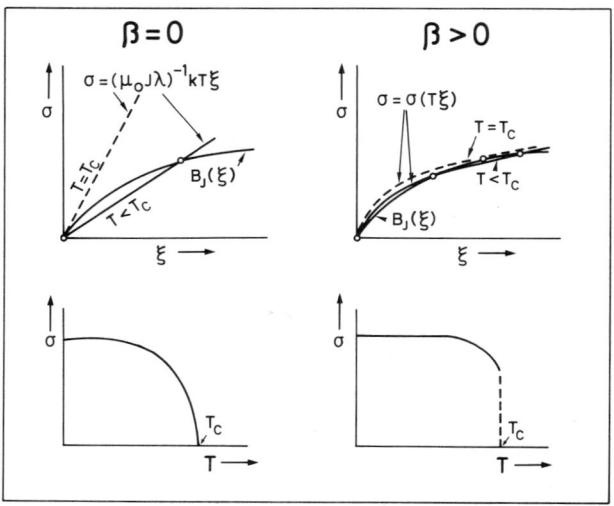

Fig. 101. Second- and first-order magnetic phase transitions in the effective field model of bilinear (Heisenberg) and biquadratic exchange. Left: only bilinear exchange. Right: with biquadratic exchange. β: biquadratic exchange parameter, ξ: effective field in units of $k_B T$. Top diagrams: graphical solution of the self-consistent equations. Bottom diagrams: magnetization σ as a function of temperature T. $\mu_0 J =$ saturation magnetic moment per ion; $\lambda =$ molecular field constant; $B_J =$ Brillouin function for spin J.

$$h' = \lambda\sigma + \beta\sigma^3, \qquad (3.1.4\text{-}2)$$

where β is proportional to a sum of the K_{ij}'s of Eq. (3.1.3-4). If β is large enough, there occur more than two solutions of Eqs. (3.1.4-1,2), as Fig. 101, top right, shows. As T increases from $T = 0$ K, the solution for finite σ may suddenly be lost at a temperature T_C (or T_N), where σ vanishes discontinuously (Fig. 101, bottom right).

Biquadratic exchange is often invoked to explain experimental magnetization vs. T data which show a steep falloff of σ near T_N or T_C.[339,340,347] However, it does not explain the moment-jump type of transition. Furthermore, the use of biquadratic exchange is artificial without a microscopic theory of its origin.

Such a theory is the *exchange striction* model of M. E. Lines, which was applied to MnO.[339] This model assumes that the magnetic exchange interactions depend strongly on the distances between the ions, so that in the

3.1. LOCALIZED ELECTRON THEORIES

magnetically-ordered state the free energy of the solid can be lowered by a distortion of the lattice. This decreases the magnetic energy by bringing certain pairs of strongly interacting ions closer together at the expense of raising the elastic energy. The effect of exchange striction was shown to be the introduction of biquadratic interactions. However, the exchange striction in actinide compounds is believed to be small.[14]

An earlier theory of a similar nature is the *Bean-Rodbell theory*,[340] in which the distortion is restricted to being an isotropic volume change and the molecular-field parameter λ is considered to depend phenomenologically on the volume. However, the volume change $\Delta V/V \cong 10^{-2}$ predicted by this theory at a first-order transition is much larger than that observed in actinide compounds (see Chapter 2).

3.1.4b. Allen's Theory of UO_2

A microscopic theory of the origin of quadrupole–quadrupole interactions via the *spin-lattice interaction* is S. J. Allen's theory of UO_2.[14] Exchange striction is assumed to be small, and the interaction of the localized uranium moments with the lattice occurs through the change in the crystal field produced by local lattice distortions, i.e., through *magneto-striction*. The crystal-field ground state is assumed to be the Γ_5 triplet of the $^3H_4$ term of the $5f^2$ configuration of the U^{4+} ion. The three levels of the triplet will be mixed by the interaction of the quadrupole moment of the Γ_5 state and the electrostatic potential created by the displacement of the eight surrounding O^{2-} ions from their equilibrium positions. Of the uniform ($k = 0$) modes of distortion, only the three shown in Fig. 102b can mix the Γ_5 levels. A distortion raises the lattice elastic energy but lowers the electronic energy of the $5f^2$ ground state. There will also be an effective interionic quadrupole–quadrupole interaction, because a displacement of the O atoms around one U ion also affects the electrostatic energy of the neighboring U ions.

In order to determine the effects of the quadrupole–quadrupole interaction on the ground state and magnetic transition of UO_2, we consider the following single-ion Hamiltonian $\mathcal{H}$:

$$\mathcal{H} = -Z_u J\langle S_z \rangle S_z - 2\epsilon \langle \mathcal{O} \rangle \mathcal{O}. \tag{3.1.4-3}$$

The first term represents the interaction between the "effective spin" ($S = 1$) of the Γ_5 triplet of a particular U^{4+} ion and the molecular field $Z_u J\langle S_z \rangle$ produced by the number Z_u of nearest neighbor U^{4+} ions, where the angular brackets denote a quantum-statistical average. The second term is the effective

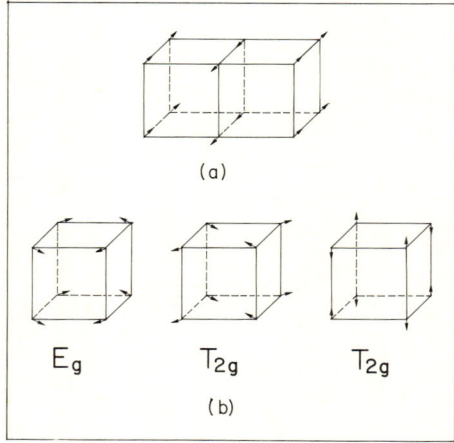

Fig. 102. Lattice distortion in UO_2. The corners of the cubes represent the positions of the oxygen atoms surrounding the uranium sites. The arrows indicate the direction of displacement. a). Observed mode (Ref. 190); b). Three modes considered in Allen's theory (Ref. 14).

quadrupole–quadrupole interaction in the molecular-field approximation. The energies J and ϵ are adjustable parameters. The form of the quadrupole operator $\mathscr{O}$ depends on the symmetry of the distortion. For the distortion considered by Allen (see Fig. 102b), one finds

$$\mathscr{O} = S_z^2 - S_x^2 = \frac{1}{2}\begin{pmatrix} 1 & 0 & -1 \\ 0 & -2 & 0 \\ -1 & 0 & 1 \end{pmatrix}, \qquad (3.1.2\text{-}4)$$

where the matrix representation of $\mathscr{O}$ is relative to three orthogonal unit vectors representing the $S_z = -1$, 0, and $+1$ states of the $S = 1$ manifold. The z axis is chosen to lie along the direction of the average value of the spin **S**, which is assumed to be the $\langle 110 \rangle$ direction. The Hamiltonian $\mathscr{H}$ becomes:

$$\mathscr{H} = \alpha\begin{pmatrix} 1 & 0 & 0 \\ 0 & 0 & 0 \\ 0 & 0 & -1 \end{pmatrix} + \beta\begin{pmatrix} 1 & 0 & -1 \\ 0 & -2 & 0 \\ -1 & 0 & 1 \end{pmatrix}, \qquad (3.1.4\text{-}5)$$

where

$$\alpha \equiv -Z_u J \langle S_z \rangle; \qquad (3.1.4\text{-}6)$$

$$\beta \equiv -\epsilon \langle \mathscr{O} \rangle. \qquad (3.1.4\text{-}7)$$

The three eigenvectors ψ and eigenvalues ω of $\mathscr{H}$ are easily found to be the following:

3.1. LOCALIZED ELECTRON THEORIES

$$\psi_0 = \begin{pmatrix} 0 \\ 1 \\ 0 \end{pmatrix}, \quad \omega_0 = -2\beta, \qquad (3.1.4\text{-}8)$$

$$\psi_\pm = \frac{1}{\sqrt{(1+v_\pm^2)}} \begin{pmatrix} v_\pm \\ 0 \\ 1 \end{pmatrix}, \quad \omega_\pm = \beta \pm \sqrt{(\alpha^2 + \beta^2)}. \qquad (3.1.4\text{-}9)$$

In these equations,

$$v_\pm = \frac{\beta}{\alpha \mp \sqrt{(\alpha^2 + \beta^2)}} \qquad (3.1.4\text{-}10)$$

We consider the case for which both $Z_u J$ and ϵ are positive numbers. At $T = 0$ in the ordered state where $\alpha \neq 0$ and $\beta \neq 0$, the only populated state is ψ_- having the energy ω_-. The average values of S_z and $\mathscr{O}$ in this state are determined by the self-consistent equations

$$\langle S_z \rangle = \langle \psi_- | S_z | \psi_- \rangle = (v_-^2 - 1)/(v_-^2 + 1), \qquad (3.1.4\text{-}11)$$

$$\langle \mathscr{O} \rangle = \langle \psi_- | \mathscr{O} | \psi_- \rangle = \tfrac{1}{2}(v_- - 1)^2/(v_-^2 + 1), \qquad (3.1.4\text{-}12)$$

where $v_\pm$ are given by Eq. (3.1.4-10). For the condition $\epsilon/Z_u J = 0.45$ assumed by Allen to apply to UO_2, the solution of Eqs. (3.1.4-11,12) is $\langle S_z \rangle = 0.96$, $\langle \mathscr{O} \rangle = 0.64$. Thus, in the ground state, the ordered spin and quadrupole moment are reduced from their maximum values of 1.0, even at $T = 0$. (Note that in Allen's second paper,[14] the quantity $\langle \mathscr{O}_{xy} \rangle$ is equal to $\tfrac{1}{2}\langle \mathscr{O} \rangle$.) The temperature dependence of $\langle S_z \rangle$ and $\langle \mathscr{O} \rangle$ is determined by the generalization of Eqs. (3.1.4-11,12) to finite temperature, which yields the behavior shown in Fig. 103, where the transition at T_N is seen to be of the first order. To explain the physical reason for the transition, we show in Fig. 104b the energy levels ω_0 and $\omega_\pm$ at $T = 0$ K for the same case ($\epsilon/Z_u J = 0.45$) considered above. By comparison to the situation in Fig. 104a, we see that the effect of the quadrupole–quadrupole interaction ϵ is to render the two upper levels ψ_0 and ψ_+ nearly degenerate ($\omega_0 \cong \omega_+$). The average values of S_z in these two states are roughly zero and -1, respectively. Thus, when the temperature T is increased to a value approaching $(\omega_0 - \omega_-)/k_B$, there occurs a rapid decrease in $\langle S_z \rangle$ and $\langle \mathscr{O} \rangle$, due to the population of the two states ψ_0 and ψ_+. The splitting between all these levels also decreases, because this splitting depends on $\langle S_z \rangle$ and $\langle \mathscr{O} \rangle$. This "bootstrap" effect causes a catastrophic collapse of the levels and a first-order transition as T is

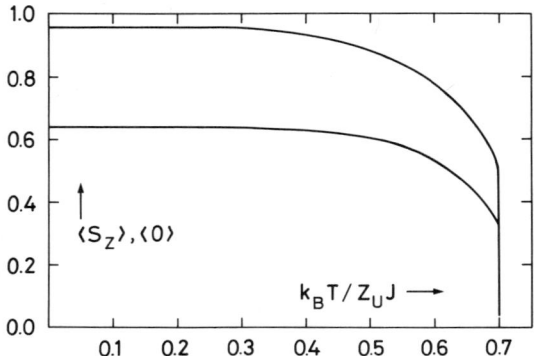

Fig. 103. Average spin $\langle S_z \rangle$ (upper curve) and quadrupole moment $\langle \mathcal{O} \rangle$ (lower curve) of the uranium ion as a function of the reduced temperature in Allen's model of UO_2. Z_u: Number of interacting neighbors. J: Exchange constant.

raised to a critical value T_N (see Fig. 103). By contrast, the transition in the case of Fig. 104a is continuous (second-order) at T_N, because the thermal decrease in $\langle S_z \rangle$ is less rapid for being "spread out" over three equally spaced energy levels.

The comparison to experiment[11] is shown in Fig. 71 where the first-order transition but not the detailed shape of the magnetization curve is explained. The experimental value[11] $1.78\,\mu_B$ for the ordered moment at $T = 0\,K$ indicates a reduction of $\cong 11\%$ from the value $2.0\,\mu_B$ expected for a pure Γ_5 ground state.

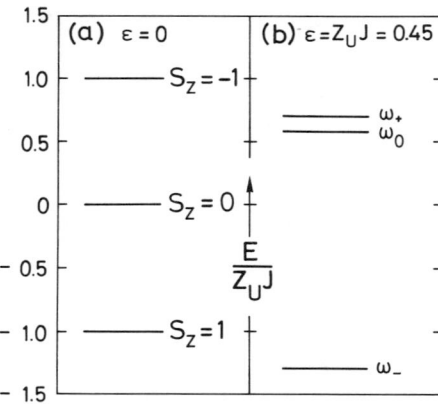

Fig. 104. Energy levels of the uranium ion at $T = 0\,K$ in the molecular field approximation in Allen's model of UO_2. a). Without quadrupole coupling; b). With quadrupole coupling. ϵ: quadrupole interaction parameter. cf. Eqs. (3.1.4–10).

3.1. LOCALIZED ELECTRON THEORIES

The latter reduction is much larger than the value of $\cong 4\%$ expected from Allen's theory. Also, the experimental paramagnetic susceptibility[195] indicates a value of the Curie–Weiss temperature θ more than twice as large as that predicted by the theory (see Table 8). The latter discrepancy may result from the fact that only the exchange interaction between first nearest-neighbor uranium ions is considered in Allen's calculation of the magnetization curve. On the other hand, the temperature dependence of the elastic constant[187] $c_{44}(T)$ is qualitatively explained by the theory.

Several other types of phase transitions are possible in the system described by Eq. (3.1.4-3), depending on the value of $f = \epsilon/Z_u J$. The possibilities are summarized in Fig. 105. Especially interesting are the possibilities (for $f > 1/2$) of two successive phase transitions and a quadrupole ordered phase ($\langle \mathcal{O}_{xy} \rangle \neq 0$) without an ordered dipole moment, i.e., $\langle S_z \rangle = 0$. Perhaps this latter possibility is realized in NpO_2, where there is no detectable ordered dipole moment,[201, 202, 203] despite all other signs of an order–disorder transition at $T = 25$ K being present (see Section 2.6).

Allen also derived the spectrum of elementary excitations in the ordered state by considering the bilinear exchange coupling $J\mathbf{S}_i \cdot \mathbf{S}_j$ and the indirect quadrupole–quadrupole interaction between U^{4+} ions. The latter interaction can be understood from the observation that a rotation of the quadrupole on

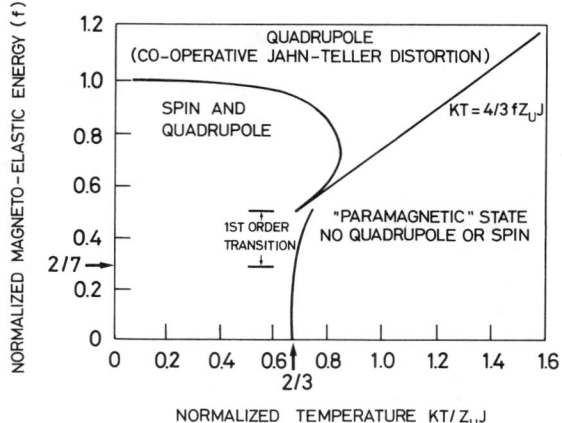

Fig. 105. Phase diagram of a model substance with exchange and distortion-induced quadrupole interactions. $f = \epsilon/Z_u J$. ϵ, J: Quadrupole and exchange interaction parameters. Z_u: Number of interacting neighbors. From Allen, Ref. 14.

site i causes a distortion of the lattice which propagates as a (virtual) phonon to site j, where the phonon interacts with the quadrupole at that site.

The excitation spectrum was determined by the usual techniques of second quantization. The exchange and quadrupole interaction parameters listed in Table 19, row A, were obtained by fitting the calculated spectrum to experimental values[25] at four points in the magnetic Brillouin zone. The quadrupole interaction so obtained is as large as the magnetic exchange ($\cong 35$ cm^{-1}). It is also noteworthy that the indirect mechanism (via virtual phonon exchange) yields a much larger interaction between the uranium quadrupoles than the purely electrostatic interaction, which is estimated to have the strength of $\cong 1$ cm^{-1}.

A shortcoming of Allen's theory is that a simultaneous fit to both the excitation spectrum and the temperature-dependent magnetic data could not be obtained with a single set of the four parameters. The discrepancies between rows A and B of Table 19 are large for the quadrupole interaction parameter C and the predicted distortions $|U_i|$. Furthermore, the predicted external strain $|U_6|$ is larger than the result $\Delta l/l < 10^{-4}$ found by Brandt and Walker.[187] The author explains this by pointing out that a bulk single crystal of UO_2 contains many magnetic domains,[25] so that the experiment measures an average of the strains over all domain orientations. An important feature of this microscopic treatment is the occurrence of large internal distortions ($u_z/r_0 \cong 10^{-2}$) without large observable external strains ($|U| \cong 10^{-4}$). Here r_0 is the equilibrium distance between neighboring O ions.

As we described earlier in Section 2.6, recent elastic neutron diffraction experiments[184,190] have detected an internal distortion in the ordered state of UO_2. The distortion, which involves a 0.014 Å displacement of the oxygen atoms from their ideal CaF_2 positions, is shown in Fig. 102 and is not one of the distortions considered by Allen. However, the magnitude and existence of such a distortion is consistent with the basic idea of this theory.

Siemann and Cooper[283] investigated the relative stability of the distortion proposed by Allen, and of that actually observed, by calculating the electrostatic and elastic energies of the two distortion models. They find that under their assumptions the Allen mode has the lower total energy, in contradiction to experimental results. They conclude that the lattice must be softer with respect to the observed mode.

Solt and Erdös[290] criticize the above conclusion on the basis that phonon data show no softness against the Allen mode. Moreover, taking into account an interaction neglected by Siemann and Cooper (namely, the interaction between the internal and external strains), Solt and Erdös find that the Allen

3.1. LOCALIZED ELECTRON THEORIES

Table 19. Theoretical Parameters and Predictions of Allen's Theory[14] of UO_2. ($\delta S_z/S_z$ is the spin deviation at $T = 0$ K. The external strains are denoted by U_i ($i = 1, 2, \ldots, 6$) in the Voigt notation,[387] and the relative internal displacement of the O^{2-} ions is $|U_z|/r_0$. θ is the Curie–Weiss constant, r_0 the equilibrium O–O distance, and the other symbols are explained in the text.)

	Parameters					Predictions[a]				
	$z_u J$	$|A|$	$|B|$	$|C|$	$\delta S_z/S_z$	$|U_1|$	$|U_3|$	$|U_6|$	$|U_z|/r_0^{-1}$	θ
(A)	35 cm^{-1}	1.5	31.5	31.0	0.063	0.5×10^{-4}	10^{-4}	3×10^{-4}	8×10^{-3}	—
(B)	32.4 cm^{-1}	0	—	16.2	0.043	0	0	3×10^{-3}	4×10^{-3}	88 K
Exp.	—	—	—	—	—	$<10^{-4}$	$<10^{-4}$	$<10^{-4}$	—	200 K

[a] Note: $|U_4| = |U_5| = 0$, $|U_1| = |U_2|$.

mode is even more favored energetically. The latter two authors show that direct electrostatic quadrupolar interactions between uranium ions favor the observed mode and, if enhanced by polarization and ligand effects, stabilize it. The required enhancement is a factor of four or five above the vacuum values of the quadrupolar interactions.

3.1.4c. Theories of NpO_2

Recall that the magnetic susceptibility and specific heat of NpO_2 show anomalies at $T_0 = 25.4$ K very similar to those seen in the corresponding data on isostructural UO_2, which orders magnetically at $T_N = 30.8$ K (see Figs. 72, 73, and 81). Neutron diffraction and Mössbauer experiments[201-203] combined yield the surprising conclusion that the Np magnetic moment, if any, is no larger than 0.01 μ_B per Np ion.

In response to the challenge of explaining the nature of the transition at $T_0 = 25.4$ K in NpO_2, the following three models have been proposed:

1. The Crystal-Field Model. Solt and Erdös[286,348,349] apply the localized crystal-field model, assuming the Np ions to be in the tetravalent $5f^3$ configuration. The value of the quantity x, which measures the ratio of the fourth- to sixth-order crystal-field parameters, is deduced from data on UO_2 to be $x = -0.74$ in NpO_2. For this value of x the ground state of the Np^{4+} ion is predicted to be a Γ_8 quartet, in agreement with experiment.[242,243]

The next crucial assumption in the model is that, as in UO_2, the quadrupolar interaction between actinide ions in NpO_2 leads to a lattice distortion setting in at $T_0 = 25.4$ K. The symmetry of the distortion is assumed to be the same as that found in UO_2 by neutron diffraction (Fig. 102). The effect of the distortion is to split the Γ_8 quartet ground state into two doublets having magnetic moments with z components μ_z which depend on the crystal field. The key result of the model is that the value of μ_z in the ground-state doublet vanishes for $x_0 = -0.75$, which is very close to the theoretical value $x = -0.74$ (see Fig. 106). Thus, in this model the transition at T_0 in NpO_2 is a quadrupolar ordering with a lattice distortion, which in combination with the crystal field produces a ground state with zero (or near zero) magnetic moment per Np ion. The assumptions of the model which are as yet unverified or unexplained are the lattice distortion and the strong anisotropy required to force the angular momentum into the (100) direction.

2. Antiferromagnetic Hole Model. Because of the uncertainties involved in the above model, a completely different approach[350] has been proposed which assumes a $5f^4$ configuration for each Np^{3+} ion. The additional electrons are

3.1. LOCALIZED ELECTRON THEORIES

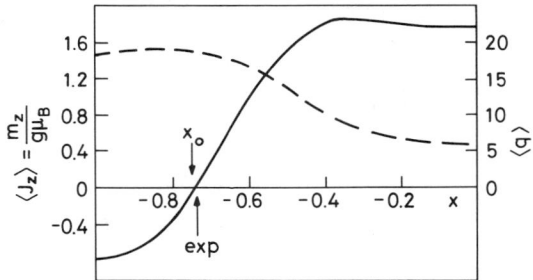

Fig. 106. Magnetic moment m_z at a Np^{4+} ion in units of $g\mu_B$ (i.e. $\langle J_z \rangle$) for one component of the ground-state doublet, as a function of the field parameter x (solid line); $g = \frac{8}{11}$. The quadrupole moment $\langle q \rangle = \langle Q_{xy} \rangle / (\frac{3}{2}\overline{r^2}\alpha)$ is also plotted (dashed line); $\alpha = \langle J\|\alpha\|J\rangle$ and $\overline{r^2} =$ radial moment of the $5f$ wave function. The arrow "exp" points to the value deduced using data on UO_2. From Ref. 349.

assumed to be missing from the oxygen $2p$ bands, leading to a deficiency of one electron, i.e., a hole located at the center of those oxygen cubes not containing a Np ion (see Fig. 107). Then the phase transition at T_0 is assumed to result from an antiferromagnetic ordering of the $s = \frac{1}{2}$ hole spins, which escape neutron detection due to their small form factor. The absence of a Np moment is explained by the $5f^4$ configuration, which allows nonmagnetic ground states. A limitation of this model is that the contribution of the Np ions to the observed magnetic susceptibility has not been calculated.

3. *Order–Disorder Transition.* The third model[286,350] also postulates a $5f^4$ Np^{3+} configuration, but differs from the hole model in that half the oxygen ions are assumed to be doubly ionized and the other half singly ionized. Above T_0 each oxygen site is occupied with equal probability by either O^{2-} or O^- ions, i.e., the time-averaged symmetry of the oxygen sublattice is simple cubic. As the temperature is lowered to $T_0 = 25.4$ K, an order–disorder transition is proposed to occur, in which below T_0 the two kinds of oxygen occupy alternate

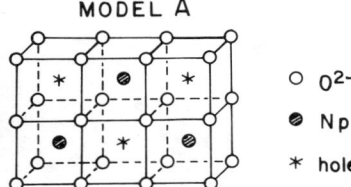

Fig. 107. Lattice structure of NpO_2, showing locations of holes in antiferromagnetic hole model of transition in NpO_2. From Ref. 350.

MODEL B

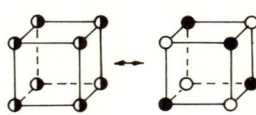

Fig. 108. Transition in NpO_2 at $T_0 = 25.4$ K according to order–disorder model. One cube of the oxygen sublattice is shown. Left: above T_0 each site is occupied with equal probability by either O^{2-} or O^- ions. Right: below T_0 the O^{2-} and O^- are arranged on alternate sites. From Ref. 350.

sites (see Fig. 108). An advantage of this model is that the magnetic susceptibility is due entirely to the Np ion, for which a Van Vleck formula for a singlet–triplet $5f^4$ system fits the data well (Fig. 109). However, there is again no experimental evidence for the existence of two kinds of oxygen ions in NpO_2.

Both the hole and order–disorder models assume the Np^{3+} ($5f^4$) configuration which not only would be unusual in the actinide dioxides but also conflicts with neutron form factor determination,[201] which favors Np^{4+}. For this reason, the latter two models have not found wide acceptance, and the crystal-field model has been preferred. However, in such a complex system as NpO_2, all alternative models should be considered until definitively ruled out by future experiments.

3.1.4d. The Long–Wang Theory of UP

The theory of the first-order moment-jump transition in UP proposed by Long and Wang[13,351] involves the assumption of quadrupole–quadrupole

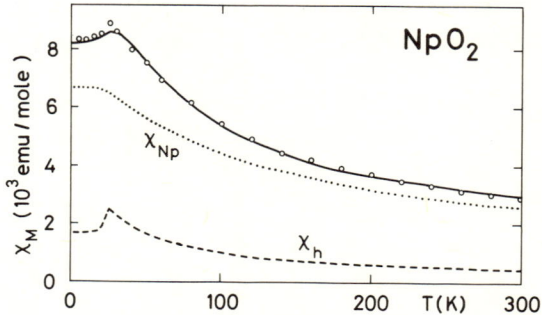

Fig. 109. Molar susceptibility χ_M of NpO_2 vs. temperature T. The circles represents the experimental points (Ref. 286). Full curve: calculated on the basis of model 3 (see text). Dotted curve: Np-susceptibility χ_{Np} obtained by subtraction of the calculated hole susceptibility χ_h. From Ref. 350.

3.1. LOCALIZED ELECTRON THEORIES

interactions between uranium ions as in Allen's model of UO_2. However, there is no microscopic derivation of the origin of the interactions, and thus no predictions of lattice distortions or anomalies in the elastic constants of UP are made. Another significant difference between the two theories is that the presence of closely spaced low-lying crystal-field levels is important in the Long–Wang theory, whereas in Allen's model only the lowest level is considered.

The Hamiltonian $\mathcal{H}_i$ of the ith magnetic ion is written in the molecular-field approximation as

$$\mathcal{H}_i = \mathcal{H}_{cf} - \mathscr{F}(o)\langle J^z\rangle J_i^z - K(o)\langle \mathcal{O}_\theta\rangle \mathcal{O}_{\theta i}, \qquad (3.1.4\text{-}14)$$

where $\mathcal{H}_{cf}$ is the crystal-field Hamiltonian appropriate for the cubic symmetry of UP (see Table 17). The quantities $\mathscr{F}(o)$ and $K(o)$ are variable parameters characterizing the isotropic and quadrupolar coupling, respectively. The second and third terms of the Hamiltonian are similar to Allen's Hamiltonian [Eq. (3.1.4-3)], but with the significant difference that the authors keep only the component $\mathcal{O}_\theta = 3(J^z)^2 - J(J+1)$ of the quadrupole operator, whereas Allen found the O_{xy} term to be dominant in UO_2. Dropping O_{xy} thus seems to be questionable.

The trivalent state for the uranium ion is assumed, and the ten states of the lowest $J = 9/2$ manifold are diagonalized numerically for each pair of trial values of $\langle J^z\rangle$ and $\langle \mathcal{O}_\theta\rangle$ and for fixed values of $\mathscr{F}(o)$, $K(o)$, and the two crystal-field parameters W and x which characterize the fourth- and sixth-order terms of the electrostatic potential acting on U^{4+}. The free energy is minimized with respect to the two ordering parameters $\langle J^z\rangle$ and $\langle \mathcal{O}_\theta\rangle$, as discussed earlier in the case of Allen's theory, to obtain the thermodynamically most stable solution at a given temperature T. The best fit to the sublattice magnetization curve of UP is obtained for the parameters $W/k_B = -40.9$ K, $x = -0.68$, $\mathscr{F}(o)/k_B = 21.9$ K, and $K(o) = 1$ K. Here, W and x are the Lea, Leask, and Wolf parameters.[312] As Fig. 110 shows, this theory gives a good fit to the NMR data near the moment-jump transition at $T = 22.5$ K, but the calculated curve does not fall off steeply enough near T_N. Also, the magnetic-powder susceptibility of UP in the ordered state could not be explained with this theory.[351]

Other limitations of the Long–Wang model are that it does not account for the sharp change in the electrical resistivity[6] of UP at the moment-jump transition or for the fact that the very similar transition at $T = 63$ K in UAs is associated with a change in the type of magnetic order.[2] This theory provides no mechanism to explain a change of the type of the magnetic order, i.e., the magnetic symmetry.

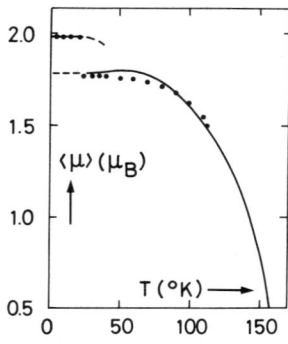

Fig. 110. Best fit of the theory (solid curve) of Long and Wang (Ref. 13) to the experimental sublattice magnetization values (solid circles) obtained by nuclear magnetic resonance experiments (Ref. 1) on UP.

3.1.4e. Blume's Level-Crossing Theory

The Hamiltonian [Eq. (3.1.1-1)] of the $5f$ electrons in a crystal contains a magnetic interaction term and a crystal-field interaction term. The theories discussed above use a nonHeisenberg-type magnetic Hamiltonian $\mathcal{H}_{\text{mag}}$ as the source of first-order transitions. It should be noted that the crystal field itself can give rise to such transitions, even in the absence of nonHeisenberg exchange. This was shown in the theory of UO_2 proposed by Blume in 1966.[352] He assumed the crystal-field ground state of the U^{4+} ion in the $5f^2$ configuration to be the Γ_4 nonmagnetic singlet, with the Γ_5 triplet lying above it by an energy Δ (see Fig. 111a). If the effective field $\lambda\sigma$ (σ = magnetization) is large enough, the lowest magnetic sublevel of the triplet can be split off and fall lower in energy than the singlet, so that at $T = 0\,K$ the solid is magnetically ordered (Fig. 111b). As T rises, the self-consistent molecular field $\lambda\sigma$ decreases, and the splitting of the triplet levels becomes smaller. As the energy of the lowest magnetic level approaches that of the singlet, the decrease of the effective field caused by ions populating the nonmagnetic singlet causes, by a bootstrap effect, a sudden collapse of the ordered state and a first-order transition at T_N (Fig. 111c).

The experimental neutron diffraction data[11] have been well fitted by this theory; but, as already discussed in Section 2.6, the weight of experimental evidence favors a Γ_5 triplet ground state for UO_2, rather than the singlet required and used by the Blume theory. Thus, the good neutron diffraction fit seems to be fortuitous. The level-crossing scheme also cannot explain the first-order moment-jump transition in UP unless an additional nonHeisenberg exchange is considered, as Long and Wang have shown.[13] However, the first-order transition at $T_N = 2.61\,K$ in UI_3[12] may be explicable as a level crossing, as discussed later. Note that a nonmagnetic crystal-field ground state is not a

3.1. LOCALIZED ELECTRON THEORIES

necessary condition for a Blume-type level crossing. If the ground state has a much smaller magnetic moment than that of the first excited state, a first-order transition may occur, as shown in Fig. 111c, because the splitting of the magnetic sublevels of the two crystal-field states in the exchange field is proportional to their magnetic moments. In practice, the application of the level-crossing theory is complicated by the presence of nonvanishing off-diagonal matrix elements of the exchange (Zeeman-type) field in the set of crystal-field states. In this case, the magnetic moment of a level is generally a function of the exchange field. In particular, the field-induced moment of the lowest level may be so large that the upper level cannot "cross" it in the manner shown in Fig. 111b. Even if the levels cross, there is often no discontinuity in the magnetic moment. In UO_2, the off-diagonal exchange matrix element between the Γ_1 and Γ_5 levels is zero, so this complication did not arise in Blume's theory.

3.1.4f. A Theory of UI_3

The crystal-field method described earlier in this section has also been applied to explain the first-order phase transition of UI_3. The crystal structure and experimental sublattice magnetization of UI_3 are shown in Figs. 85 and 89,

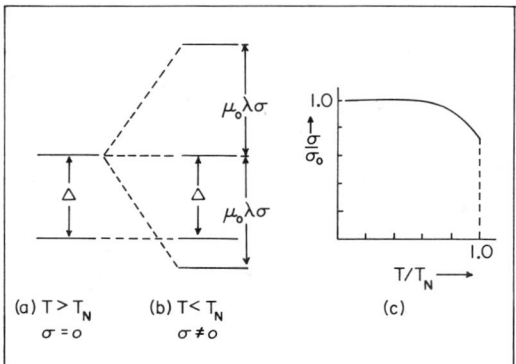

Fig. 111. Illustration of Blume's level-crossing theory (Ref. 352) of first-order magnetic phase transitions. Level scheme: a). in the paramagnetic state. b). In the magnetically ordered state. c). relative sublattice magnetization *vs.* reduced temperature. The symbols are explained in the text.

Chapter 2. The self-consistent molecular field λ is added to the crystal-field Hamiltonian $\mathcal{H}_{cf}$, given by

$$\mathcal{H}_{cf} = B_2[Y_2^0 + a(Y_2^2 + Y_2^{-2})] + B_4[Y_4^0 + b(Y_4^2 + Y_4^{-2}) + c(Y_4^4 + Y_4^{-4})]$$
$$+ B_6[Y_6^0 + d(Y_6^2 + Y_6^{-2}) + e(Y_6^4 + Y_6^{-4}) + f(Y_6^6 + Y_6^{-6})]. \quad (3.1.4\text{-}15)$$

Although the point symmetry C_{2v} of the U ion allows for arbitrary values of all coefficients in $\mathcal{H}_{cf}$, for simplicity of the numerical calculation it was assumed that $a, b, c, d, e,$ and f have their values as given by a point-charge model including charges up to a distance of 60 Å from the ion; B_2, B_4, and B_6 were varied freely. This is to some extent justified, since linear shielding modifies only B_k, but not the coefficients $a, \ldots, f$. (See Section 3.1.2-4, shielding.) The $^4I_{9/2}$ multiplet of the $5f^3$ configuration of U^{3+} in the orthorhombic lattice of UI_3 breaks up into 5 magnetic doublets having magnetic moments which vary strongly with the crystal-field parameters and with the effective field.

The calculated values of the parameters do not yield a first-order transition, but an excellent fit to the sublattice magnetization (see Fig. 89) is obtained for a certain choice of parameters. To explain the first-order transition it was necessary to vary the coefficients $a = B_2^2/B_2^0$ in the Hamiltonian $\mathcal{H}_{cf}$ (Eq. 3.1.4-15).

The paramagnetic susceptibility $\chi(T)$ of UI_3 calculated with the same values of the parameters used to fit the NMR data does not agree well with experiment, as Fig. 88 shows. The difficulty is in obtaining the correct Néel temperature T_N. One can obtain a simultaneous fit to $\chi(T)$ and the relative sublattice magnetization (σ vs. T/T_N), but the resulting prediction for T_N is 0.6 K as compared to the experimental value $T_N = 2.61$ K. Thus, it is clear that this simplified crystal-field model is not adequate to explain the properties of UI_3. The above calculations have not taken into account the inequivalence of the three sites of the U^{3+} ions in UI_3. This seems justified, because the same σ vs. T curve is found experimentally for each of the sites.

It would be interesting to know whether a fully parameterized crystal-field calculation were able to bring about a simultaneous fit of $\chi(T)$ and T_N. If not, single-ion anisotropy and possibly quadrupole–quadrupole interactions have to be included in an adequate theory of UI_3.

3.2. Theories Involving Itinerant Electrons

In the introduction to Section 3.1. we pointed out that the crystal-field model is applicable in the case where the $5f$ electronic energy levels are narrow (i.e., atomiclike) and well separated in energy from the band(s) of itinerant $6d$

3.2. THEORIES INVOLVING ITINERANT ELECTRONS

and 7s states (see Fig. 93). The weight of theoretical and experimental evidence shows that this model does not apply to the lighter actinide metals (Th, U, Np, and Pu)[353-356] and probably does not apply to some metallic NaCl-type actinide compounds[357] and intermetallics.[18,358] The crystal-field model seems to apply well to the heavy actinide metals (Am, Cm, Bk, Cf) and the oxides and halides of actinides. In this section we review the theories which take account of the partly itinerant (bandlike) character of the $5f$ electrons and their proximity to the $6d-7s$ band.

3.2.1. Band Calculations

3.2.1a. Introduction

In the band model the electrons in the crystal are assumed to move in an average potential due to the atomic nuclei and the other electrons. The resulting eigenstates are independent, single-particle band states labeled with a band index and a wavevector **k** in the Brillouin zone. The intra-atomic Coulomb correlations between the electrons are neglected, and thus the band picture is at the opposite extreme from the crystal-field model, in which the quantum states are highly correlated, atomiclike, localized, many-particle states. The actinide $6d$ and $7s$ electrons are best treated in the band model because of the large overlap of d and s atomic orbitals centered on neighboring sites,[355,356] but the correct description of the $5f$ electrons may not correspond to either the band or crystal-field model.

In the case of the lighter actinide metals, the absence of local magnetic moments indicates that the $5f$ electrons may be itinerant. The bandlike character arises not only from the f-f overlap integral T_{ff}, given by

$$T_{ff} = \int d^3r \phi_{fi}^*(\mathbf{r}) v_i(\mathbf{r}) \phi_{fj}(\mathbf{r}) \qquad (3.2.1\text{-}1)$$

for $5f$ orbitals ϕ_{fi}, ϕ_{fj} on adjacent sites i and j, but also from overlap T_{df} or *hybridization* with the $6d$ (or $7s$) atomic or Wannier orbitals:

$$T_{df} = \int d^3r \phi_{di}^*(\mathbf{r}) v_i(\mathbf{r}) \phi_{fj}(\mathbf{r}). \qquad (3.2.1\text{-}2)$$

Here $v_i(\mathbf{r})$ is the difference between the periodic potential of the metal and that of a free atom at site i.[354]

3.2.1b. Actinide Metals

Freeman and Koelling have carried out extensive band calculations on the actinide elemental metals using the relativistic augmented plane wave (RAPW) method.[355,356] In this method, the crystal potential is assumed to be spherically symmetric within a sphere centered on each atomic site and constant or nearly constant between the spheres (so-called "muffin-tin" potential). The potential in each sphere is the sum of the Coulombic atomic potentials calculated for a well-defined configuration of each atom and the "exchange potential" V_{ex}, which in Slater's approximation is given by

$$V_{ex}(\mathbf{r}) = -6\alpha \left[\frac{3}{8\pi}\rho(\mathbf{r})\right]^{1/3} \quad (3.2.1\text{-}3)$$

where $\rho(\mathbf{r})$ is the electronic density and α is a variable parameter normally equal to 2/3 or 1. Schrödinger's Equation is solved inside each sphere and the solution is matched to plane waves $e^{i\mathbf{k}\cdot\mathbf{r}}$ in the region between the spheres. Relativistic terms such as the spin–orbit coupling were also considered and found to be extremely important in determining the ordering and splitting of the bands in the actinides. The relativistic $5f$ wavefunctions are about 20% more delocalized than those calculated nonrelativistically.[356]

The methods and results of the calculations of Freeman and Koelling are reviewed in detail elsewhere;[356] we present only a few significant conclusions. In Fig. 112 are shown the energy bands of *bcc* uranium (Fig. 112a) and *fcc* americium (Fig. 112b) along one direction in their respective Brillouin zones. In both cases, the broad $7s$ bands are lowest in energy. The $5f$ bands come next, but their characteristics are strikingly different in the two metals. In uranium the $5f$ bands are about 1 eV broad and strongly hybridized with the $6d$ band. In americium, on the other hand, the $5f$ bands are extremely narrow (atomiclike) and show no evidence of hybridization, even though they intersect the $7s$ band of itinerant states. The results for Th, Np, and Pu are similar to those for U in that the $5f$ electrons are delocalized or itinerant, so that a crystal-field model is completely inapplicable. In the heavier actinides Am, Cm, and Bk the $5f$ electrons are localized, but a crystal-field description may still have to be modified by the presence of overlapping itinerant $7s$ and $6d$ states into which the localized $5f$ electrons can make thermal transitions. One should remember, however, that band theory is not reliable for the prediction of the energies of localized states (see Section 3.1).

The actinide metals should be contrasted to the rare earths in which the $4f$ electrons are quite well localized and (with the exception of Ce) energetically separated from the band states.

3.2. THEORIES INVOLVING ITINERANT ELECTRONS

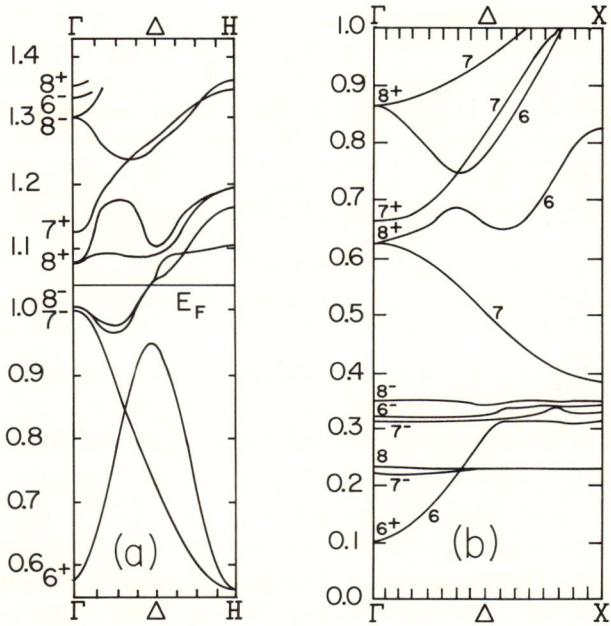

Fig. 112. Energy (in Ry) bands of (a) body-centered cubic uranium, and (b) face-centered cubic americium, as functions of the wave number along selected directions in the Brillouin zone. For U, the ionic configuration is taken as $f^3 d^1 s^2$, and Slater's factor $\alpha = 2/3$. For Am, the configuration is $f^6 d^1 s^2$, $\alpha = 1$. Results calculated by Koelling and Freeman, where the designation of levels is also explained, Refs. 355, 356.

Jullien et al.[353,354] have considered the effects of the d-f hybridization on the localization of the $5f$ electrons in the actinide metals. In the first paper,[353] both the f and d states are assumed to be broadened in energy only through hybridization with the broad $7s$ band; i.e., the overlap integrals T_{dd} and T_{ff} [Eqs. (3.2.1-1,2)] are neglected. This conflicts with the results of Freeman and Koelling,[355,356] which predict that the $6d$ and $5f$ bands in the lighter actinide metals are broadened mainly through the effect of the overlap integrals T_{dd} and T_{ff}. Jullien et al. explain the absence of localized magnetic moments in the lighter actinides as due to a large d-f hybridization of about 2 eV.

In a second paper,[354] Jullien and Coqblin improve their previous model by taking into account the d-d and f-f nearest-neighbor overlap in the tight-binding approximation. Their Hamiltonian is given by

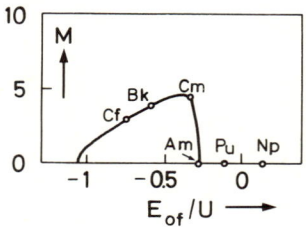

Fig. 113. The magnetic moment M in Bohr magnetons vs. E_{of}/U in the actinide series, calculated by Jullien and Coqblin, Ref. 354. E_{of} is the energy of the center of the f band; U is the intraatomic Coulomb repulsion energy between electrons of opposite spin.

$$\mathcal{H} = \sum_{k,\sigma} E_d(\mathbf{k}) c_{dk\sigma}^+ c_{dk\sigma} + \sum_{k,\sigma} E_f(\mathbf{k}) c_{fk\sigma}^+ c_{fk\sigma} + \sum_{k,\sigma} (V_{df}(\mathbf{k}) c_{dk\sigma}^+ c_{fk\sigma} + h.c.)$$

$$+ U \sum_i (n_{di\uparrow} n_{di\downarrow} + n_{fi\uparrow} n_{fi\downarrow} + n_{di\uparrow} n_{fi\downarrow} + n_{di\downarrow} n_{fi\uparrow}) \qquad (3.2.1\text{-}4)$$

$E_d(\mathbf{k})$ and $E_f(\mathbf{k})$ are energies in the d and f bands, and $V_{df}(\mathbf{k})$ is the k-dependent d-f hybridization proportional to T_{df} [Eq. (3.2.1-2)]. The Coulomb repulsion between two electrons of opposite spin on the same lattice site i is denoted by U. The operator $c_{dk\sigma}^+$ ($c_{fk\sigma}^+$) creates a d (f) electron in the band state of wavevector $\mathbf{k}$ and spin σ, and $n_{di\downarrow}$ ($n_{fi\downarrow}$) is the number operator for spin-down $d(f)$ electrons at the site i. The Hamiltonian is treated in the Hartree–Fock approximation, yielding the quantum average values of the number of electrons of each band, wavevector, and spin per actinide atom as functions of the parameters U, the band halfwidths Γ_d and Γ_f, the band centers E_{od} and E_{of}, and the hybridization $V_{df}(\mathbf{k})$. The value E_{od} is held fixed, and the f band center E_{of} is varied to fit the total number N of d and f electrons for the various actinide metals. Fig. 113 shows the local magnetic moment M as functions of E_{of}/U for the parameters $U = 1$ eV, $\Gamma_d = 2$ eV, $\Gamma_f = 0.2$ eV, $U = 1$ eV, $E_{od} = 0.7$ eV, and $V_{df}(\mathbf{k}) = 0.7$ eV independent of $\mathbf{k}$. The model accounts for the appearance of local moments in Cf, Bk, and Cm, but the assumption of a constant f band width disagrees with the calculations of Freeman and Koelling, who find a marked narrowing of the f bands along the series from U to Am. Also, in the band model there is little justification for identifying quantities like $\langle n_{d\uparrow} \rangle - \langle n_{d\downarrow} \rangle$ with a local magnetic moment.

3.2.1c. NaCl-Type Actinide Compounds

Calculations of the band structure of the uranium compounds UC, UN, UP, UAs, US, and USe were carried out by H. L. Davis[357] using nonrelativistic energy band theory and the "muffin-tin" potential approximation described in 3.2.1b. The results for paramagnetic UP are shown in Fig. 114 and resemble

3.2. THEORIES INVOLVING ITINERANT ELECTRONS

the calculations of Fig. 112a for uranium metal in that broad f bands and considerable f-d hybridization are indicated. A similar model was proposed by Bates and Unstead.[47] However, this model is difficult to reconcile with the large values of the magnetic moments ($\cong 3\,\mu_B$) found experimentally in these UX compounds and the neutron diffraction results, which point to localized $5f$ magnetic moments.[72,73] In the case of uranium metal, such broad f bands are thought to explain the absence of localized magnetic moments. Among the UX compounds, only in the case of UN is there significant evidence favoring band or itinerant antiferromagnetism.[46] The evidence is mainly the low value ($\cong 0.7\,\mu_B$) of the ordered moment and the small specific-heat anomaly at T_N (Section 2.2.4). Delocalization of the $5f$ electrons in UN may result from the smallness of the lattice constant $a_0 = 4.89$ Å, which is nearly 1 Å smaller than those of other uranium monopnictides (see Table 3). On the other hand, the neutron diffraction experiment,[53] the Curie–Weiss behavior of the paramagnetic susceptibility,[66] and the size ($3.1\,\mu_B$) of the paramagnetic moment[19] are best explained by the localized model. The specific-heat coefficient γ calculated by Davis for UN using his band results is $\gamma = 15.1$ mJ/mole-K^2, a result quite a bit smaller than the experimental value $\gamma = 49$ mJ/mole-K^2. The experiment indicates a $5f$ band intersecting the Fermi level as predicted by the theory, but the width of the $5f$ band appears to be much less than the theoretical band width.

A schematic illustration of the electronic density of states in the NaCl-type actinides proposed by Davis is shown in Fig. 115 and compared to the theory of Grunzweig-Genossar et al.,[19] who place the $5f$ levels in an energy gap between the $6d$-$7s$ band and the valence (p) band. Also shown for comparison

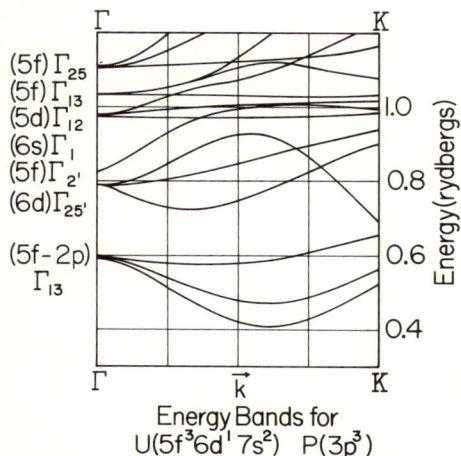

Fig. 114. Energy bands of UP as a function of the electron wave number along a selected direction of the Brillouin zone. Nonrelativistic calculation of Davis, Ref. 357. The Γ-symbols refer to the symmetry properties of each band.

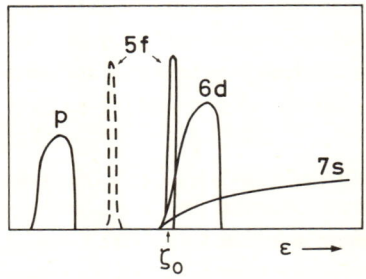

Fig. 115. Schematic illustration of the electronic density of states $D(\epsilon)$ in the NaCl-type actinide compounds, as proposed by Davis, Ref. 357 ($5f$ band traced in full line) and Grunzweig-Genossar, Ref. 19 (broken line). ζ_0: Fermi-level.

in Fig. 94 is the density of states of UAs proposed by Schönes[272] on the basis of optical experiments. The schemes proposed for other UX compounds are very similar. The proximity of the $5f$ levels to the Fermi level is clearly indicated by the high experimental γ values, numerous photoemission data,[84-86, 269-271, 299] the temperature dependence of galvanomagnetic effect in US, PuC, PuP, and PuS, and the sharp maxima in γ and electrical resistivity as functions of x in $UP_{1-x}S_x$ and $U_xTh_{1-x}S$ (see Section 2.2.9). Fisk and Coles,[45] Adachi and Imoto,[115] and Wedgwood[72] have also proposed a model of highly correlated, virtually bound $5f$ levels intersecting the Fermi surface. Again, a weak hybridization and $5f$ bandwidths narrower than those predicted in the band calculation are suggested by these authors.

In an effort to refine the calculations of Davis, the authors Allen and Brooks[306] and Brooks and Glötzel[359] considered the Darwin and mass–velocity relativistic effects which come from the Dirac equation, but neglected spin–orbit coupling. An interesting result of their calculation is shown in Fig. 95, which depicts how the width of the unhybridized f band decreases with increasing lattice constant in the uranium pnictides. Note that the f width in UN is about twice that of the other pnictides, a feature which is consistent with the majority opinion of UN as an itinerant f-electron antiferromagnet. The calculated energy bands are generally similar to those derived by Davis, but the p band is moved up in energy and hybridizes with the f states. Also, the f-d bands are more narrow, resulting in higher theoretical densities of states at the Fermi level. These features are in rough agreement with the density of states proposed by Schönes (Fig. 94) on the basis of optical measurements.

Brooks[360] has pointed out that the p-f hybridization predicted by the band calculations may explain the magnetic anisotropy observed experimentally in actinide pnictides. Brooks' explanation rests on the fact that there are two triplets of one-electron f orbitals in cubic symmetry, one of which is prolate and one of which is oblate. The oblate Γ_{25} orbitals will tend to align themselves

3.2. THEORIES INVOLVING ITINERANT ELECTRONS

in (100) planes, because the charge densities can then overlap with the pnictogen p states, and the $5f$ magnetic moments will be perpendicular to the planes. This is the configuration observed in the ordered states of the UX antiferromagnets (see Section 2.2). The above explanation of the anisotropy resembles closely the model of anisotropic exchange proposed by Cooper and Siemann[346] and is subject to the same reservations as expressed in Section 3.1.3f.

3.2.2. Spin-Fluctuation Models

The electrical resistivity and magnetic susceptibility as functions of temperature of actinide metals and intermetallics (see 2.4) have been interpreted on the basis of several "localized spin fluctuation" models.[148,149,361,358,362] In these theories, the $5f$ electronic states have a character intermediate between localized and itinerant. The hybridization and overlap interactions are assumed to prevent the formation of stable local moments and magnetic ordering, but the narrowness of the $5f$ bands and the presence of exchange interactions lead to thermodynamic fluctuations described as localized "spin fluctuations" or "paramagnons."[358] The latter excitations in the system of $5f$ electrons may be thought of as spin waves having short lifetimes and limited spatial extensions. The conduction electrons scatter from these fluctuations at finite temperatures, giving rise to a temperature dependence of the electrical resistivity $\rho(T)$ which resembles the spin-disorder scattering observed in normal magnetically ordered materials. A T^2 dependence of ρ at low temperatures is predicted by the theories and is used as the main criterion in experimentally determining if a material is, or is not, a spin-fluctuation system.

Rivier and Zlatić[362] have calculated also T, $\ln T$, and T^{-1} dependencies for $\rho(T)$ in various temperature ranges. A characteristic parameter T_{SF}, the spin fluctuation temperature, is supposed to determine the $\rho(T)$ curve, and there are six different ways of estimating T_{SF} from the experimental data, depending on which temperature range is used. However, M. B. Brodsky found that the values of T_{SF} so determined from the $\rho(T)$ data on a particular actinide intermetallic are widely different[361] depending on the method of estimation used. For example, T_{SF} = 50, 310, 192, 94, 29, or 200 K is obtained for UGa_3. Thus, the concept of the T_{SF} has little value in describing these materials and seems in general to be a vaguely defined quantity in the literature. The theory of Rivier and Zlatić shows qualitatively how the peculiar temperature dependence of $\rho(T)$ may arise through electron scattering on fluctuations.

Brodsky[361] discusses the data using the virtual bound state model of Friedel and Anderson.[363,364] The parameters of the model are U, the Coulomb repulsion between two f electrons on the same site, V_{kf}, the f-d (or f-s) hybridization matrix elements, and Δ, given by

$$\Delta = \pi \langle V_{kf}^2 \rangle n_c(E_f). \qquad (3.2.2\text{-}1)$$

In Eq. (3.2.2-1), $n_c(E_f)$ is the conduction electron density of states at the Fermi level, and, if E_k denotes an energy in the band,

$$\langle V_{kf}^2 \rangle = \sum_k |V_{kf}|^2 \delta(E_f - E_k). \qquad (3.2.2\text{-}2)$$

If $U/\pi\Delta > 1$, this model predicts local moment formation and magnetism, whereas the $5f$ electrons are delocalized if $U/\pi\Delta < 1$. The spin-fluctuation systems are then characterized by the condition $U/\pi\Delta \cong 1$. The magnetic susceptibility χ of the f electrons is given by

$$\chi = S\chi_0 = [1 - Un_f(E_f)]^{-1}\chi_0, \qquad (3.2.2\text{-}3)$$

where S is the enhancement factor, and χ_0 is the Pauli susceptibility of non-interacting $5f$ electrons (i.e., for $U = 0$). The experimental specific heat and susceptibility data are used to calculate the parameters for various actinide systems, and selected results of the calculation by Brodsky[361] are summarized in Table 20 and compared to the theories of Jullien et al.[354,358] for Np and Pu metals. The hybridization matrix elements V_{kf} and the magnitude U of the Coulomb interaction predicted by the various theories are seen to differ greatly. No calculations of the $\rho(T)$ curves are made in Brodsky's treatment.

A theory of the spin-fluctuation resistivity which fits the experimental data on Np and Pu very well (see Fig. 116) has been proposed by Jullien et al.[358] They neglect the d-f hybridization and treat the f electrons as belonging to a

Table 20. Parameters of Several Theories of Actinide Spin-Fluctuation Systems (The definition of the parameters is given in Section 3.2.2.)

	Theory (Ref.)	$N_f(E)$ (eV)$^{-1}$	U (eV)	T_f (K)	V_{kf} (eV)	S	ξ	Δ (eV)
α-Np	358		—	750	0.0	10.0	0.5	—
	354	—	1.0	—	0.7	—	—	
	361	6.0	0.110	—	0.150	3.0	—	0.071
α-Pu	358		—	280	0.0	10.0	0.4	—
	354	—	1.0	—	0.7	—	—	
	361	10.6	0.036	—	0.154	1.6	—	0.074
UAl$_2$	361	35.6	0.02	—	0.165	3.5	—	0.0086
USn$_3$	361	72	0.0107	—	0.037	4.3	—	0.0043

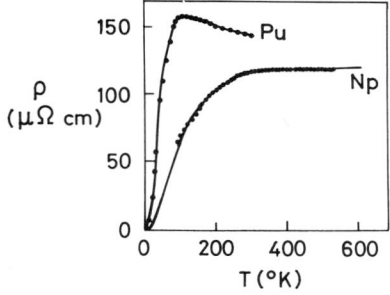

Fig. 116. Electrical resistivities of neptunium and plutonium as functions of the temperature T. The experimental data of Meaden, Ref. 389, and Olsen and Elliott, Ref. 390, are fitted by the theory of Jullien et al., Ref. 358.

narrow band of interacting itinerant particles. A curious feature of this theory is that two separate Fermi temperatures are assumed — a strongly temperature-dependent one for the f band and a constant Fermi temperature for the broad conduction band. The expression of Kaiser and Doniach[149] for the resistivity due to conduction electron scattering from the paramagnons of the f band is used to calculate $\rho(T)$. The parameters involved in fitting the data (see Table 20) are the Stoner enhancement factor S [see Eq. (3.2.2-3)], the ratio ξ of the moduli of the Fermi vectors of the conduction and f bands at $T = 0\,\text{K}$, the Fermi temperature T_f of the f band, and the high-temperature limit ρ_∞ of $\rho(T)$. The agreement with experiment is certainly striking, but the very small value $T_f = 280\,\text{K}$ found for Pu indicates extremely narrow bands for the f electrons. In such a case, there is serious doubt that the f electrons would form a well-defined Fermi surface, so that the band model would not apply. The theory of Jullien et al. requires very narrow bands in order that the temperature dependence of the Fermi level and (therefore) of the conduction electron scattering be strong enough to fit the experimental $\rho(T)$ curves.

Arko, Brodsky, and Nellis[18] have suggested that the $5f$ electrons are itinerant and form a well-defined Fermi surface only at low temperatures, and that the observed $\rho(T)$ curves reflect a gradual transition to mostly localized behavior of these electrons at higher temperature. This "thermal dehybridization" is also the key of the model of Doniach[365] which has been successfully applied to $PuAl_2$.

3.2.3. Electron Delocalization Model

A theory of the first-order magnetic moment-jump phase transitions observed in NaCl-type metallic actinide compounds and their solid solutions has been proposed by the present authors.[16,366] In this model, the energy levels of the 5f electrons are assumed to overlap those of the 6d-7s band and to lie close to or at the Fermi level (see Fig. 115). This picture conforms with the experimental and theoretical findings described in the preceding sections of this book. The 5f levels are considered to be sufficiently narrow and the hybridization sufficiently weak so that a localized model is adequate for their description. However, because of the proximity in energy of the band, thermal excitation of f electrons from the localized states to the itinerant states must be considered. Because a "delocalization" of a 5f electron to the band leaves behind an ion of higher valence, the crystal-field assumption of a unique, well-defined electronic configuration (i.e., $5f^3$) for the magnetic ions is not valid, and only the time-averaged electronic occupation of the ionic levels has significance. The Coulomb field of the fixed actinide ions as experienced by the conduction electrons is also changed by the delocalization, which leaves behind an electron "hole" in the lattice. The relative energies of the band and localized states therefore become dependent on their electronic occupation. For this reason, thermal delocalization (or localization) of electrons shifts the energies of the localized and itinerant states relative to each other and to the Fermi level, which in turn leads to further delocalization (or localization). This self-consistent or "bootstrap" effect can cause first-order transitions characterized by discontinuous changes in the ionic configuration and therefore in the localized magnetic moment as functions of temperature.

In the theory of Refs. 16 and 366, the following additional assumptions are made:

a. Each actinide site can be occupied by either M or $(M-1)$ localized f electrons. For example, in the case of UP these are assumed to correspond to the $5f^3$ and $5f^2$ configurations.

b. For each of the two ionic configurations, only the crystal-field level of lowest energy is considered. These levels are assumed to be the triplet Γ_5 and the doublet Γ_6 in the case of $5f^2$ and $5f^3$, respectively (Fig. 117).

c. There is a band of conduction electron states, derived mostly from 6d and 7s actinide orbitals and having a density of states $\mathscr{D}(\epsilon)$.

d. There is a Coulomb interaction G between an f electron localized on an ion and a band electron in a Wannier orbital centered on the site of the same ion. This is the short-range approximation made by Falicov et al.[367-371] in their theories of metal-insulator transitions.

In order to illustrate the essential mechanism of the first-order transition, let us consider the low-temperature magnetically-ordered phase of a UX-type compound in which a strong exchange field lifts the magnetic degeneracy of the Γ_5 and Γ_6 manifolds. Let Δ' be the energy required to add an f electron to

3.2. THEORIES INVOLVING ITINERANT ELECTRONS

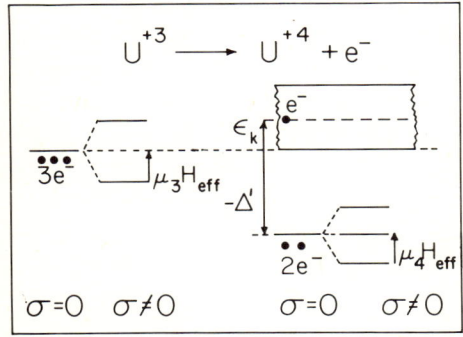

Fig. 117. Electron delocalization transition in actinide compounds, e.g., $U^{3+}(f^3) \leftrightarrow U^{4+}(f^2)$ + conduction electron. Δ': energy difference between $5f^3$ and $5f^2$ configurations. ϵ_k: energy of a conduction electron. μ_i: magnetic moment of U^i ion. σ: sublattice magnetization. H_{eff}: effective magnetic field.

a $5f^2$ ion and make the $5f^3$ configuration. We choose the zero of one-electron energies to correspond to the bottom of the conduction band, as shown in Fig. 117.

Let us now investigate the stability of the *localized phase* in which all actinide ions have the $5f^3$ configuration and (for simplicity) the conduction band is empty. The energy E_0 required to promote one electron to the bottom of the band is

$$E_0 = 3G - \Delta' \qquad (3.2.3\text{-}1)$$

where the term $3G$ represents the average value of the short-range Coulomb repulsion between the delocalized electron and the $(N-1)$ ions having the $5f^3$ configuration. Since the probability that a band electron is found in a given atomic cell is $\sim N^{-1}$, one may neglect the interaction with the single $5f^2$ "hole." If E_0 is non-negative, the localized phase is stable against single-electron excitations. Next we consider the *delocalized* phase, in which all the N actinide ions have the $5f^2$ configuration and N electrons fill the band up to the Fermi energy ζ_1. The energy E_1, required to remove one electron from the band at the Fermi level and localize it in the f shell of one of the ions, is

$$E_1 = \Delta' - \zeta_1 - 2G. \qquad (3.2.3\text{-}2)$$

If the parameters Δ', G, and ζ_1 are such that both the energies E_0 and E_1 are non-negative, then both the localized and delocalized phases are stable against excitations and are possible physical states of the solid. At $T = 0$ K, the ground state of the solid is the phase of lowest internal energy U. In the theory of Ref. 16 the ground state is the delocalized phase. However, the excitation energy E_0 of the localized phase is much smaller than that of the delocalized phase for the values of the parameters appropriate for the actinide compounds. If the solid could "switch over" to the localized phase, its entropy S would be

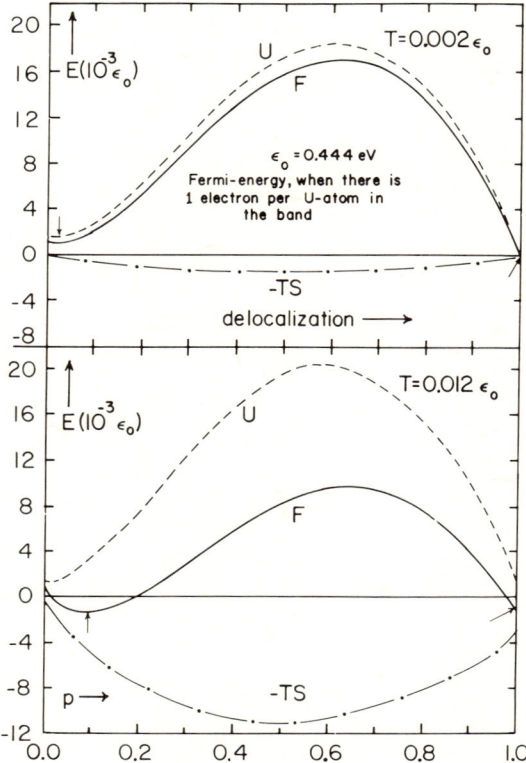

Fig. 118. Internal energy U, the quantity $-TS$ ($S =$ entropy), and free energy F of uranium phosphide UP as a function of the average number p of U^{4+} ions per U ion, for two different temperatures. ϵ_0 is the Fermi energy, T is in units of k_B. The arrows indicate minima of F. From Ref. 16.

increased because of the thermal excitation of electrons across the small gap E_0 from the $U^{3+}(5f^3)$ ions to the conduction band. This thermal promotion of electrons leaves behind a random distribution of $5f^2$ "holes" in the lattice and thus increases the entropy. For this reason, even though the delocalized phase has the smaller internal energy at $T = 0$ K, it becomes thermodynamically favorable for the solid to change over to the localized phase at a higher temperature T'.

In Fig. 118 one can visualize the first-order magnetic phase transition with the example of UP. Plotted are the internal energy U, the product of T and the entropy S, and the free energy F, as functions of the average number p of U^{4+}

3.2. THEORIES INVOLVING ITINERANT ELECTRONS 161

ions, per U ion. Hence p is the average number of band electrons per U ion produced by delocalization from a U^{3+} ion. In the top curve, the temperature (in units of the Fermi energy ϵ_0) is low, and the absolute minimum of F is at $p = 1$. At a higher temperature (bottom diagram) the relative minimum indicated in the upper diagram by an arrow close to $p = 0$ becomes the absolute minimum. Hence at a temperature intermediate between the two shown, a phase transition from $p = 1$ to $p = 0.1$ takes place. This is the physical explanation of the origin of the first-order electron localization transition predicted by the theory of Ref. 16. At the latter transition, the ordered magnetic moment per actinide ion changes suddenly due to its different values in the two ionic configurations (Fig. 117) yielding the characteristic "moment-jump" transition seen in Figs. 6 and 14.

The electron delocalization model has had the following successes:

1. The theory produces a detailed explanation of the sublattice magnetization of UP (Fig. 6a), UAs (Fig. 14a), and NpC (Fig. 30a) including the first-order "moment-jump" transitions, as functions of the temperature. In the case of UP, the theory predicted that the transition at T_N is also first-order, a prediction which was later verified by sensitive thermal-expansion measurements.[253]

2. The theory explains why the "moment-jump" transitions at $T_C = 220\,\text{K}$ in NpC[4] and $T' = 63\,\text{K}$ in UAs[2] are accompanied by a change in the type of magnetic ordering (i.e., the magnetic symmetry) whereas the otherwise similar transition at $T' = 22.5\,\text{K}$ in UP is not associated with such a change.[53] The explanation is based on the assumed existence of RKKY-type exchange interactions. As can be seen from Fig. 99, the most stable type of magnetic ordering depends on the number w of conduction electrons. Thus, at an electron localization (or "moment-jump") transition, where there occurs a decrease in w, the magnetic symmetry may or may not change, depending upon the number of conduction electrons just below and just above the transition temperature.

3. The temperature dependence of the magnetic powder susceptibility χ in the ordered state of UP[5] and UAs[6] is quantitatively explained, as shown in Figs. 6b and 14b.

4. The calculated latent heats $T\Delta S$ of the transitions at T' and T_N in UAs and UP agree well with experiment.

5. The theoretically predicted decrease in the number of conduction electrons in UP and NpC (see Ref. 16) at the "moment-jump" temperature correlates well with the observed increase in the electrical resistivity ρ of these compounds (Figs. 11 and 30b).

6. The magnetic phase diagrams of $UAs_{1-x}P_x$ (Fig. 40), $UP_{1-x}S_x$ (Fig. 41),

and $UAs_{1-x}S_x$ (Fig. 44) as functions of composition x and temperature T have been calculated by the theory and agree reasonably well with experiment. The calculation is made by assuming that the theoretical parameters vary linearly with x between the values deduced in Ref. 10 for UP, UAs, and US. A transition, such as that seen[10] in $UP_{0.75}S_{0.25}$, Fig. 42, involving a change in the magnetic symmetry at a certain temperature T'' without a discontinuity in the ordered moment, also finds an explanation. In this model, such a transition results from a thermal delocalization of electrons across the gap E_0 (Eq. 3.2.3-1) in the localized phase. The delocalization causes a continuous increase in the number w of band electrons per ion as a function of temperature above $T = T'$. If w reaches a critical value separating the regions of stability of two magnetic phases such as the AFM-I and FM phases shown in Fig. 99, a transition between the two phases is predicted to occur. No discontinuity in the ordered moment per ion is expected in the theory at the latter type of transition, because there is no discontinuity in the average valence of the actinide ions or in the effective molecular field.

J. Grunzweig-Genossar[372] has proposed a quite different theory of the magnetic symmetry changes based on a RKKY model in which w remains constant. The transitions in this theory result from the temperature-dependent perturbation of the Fermi surface by the effective field of the array of localized magnetic moments (see Section 3.1.3e). However, the latter theory cannot explain the moment-jump transition in UP, where there is no change in the magnetic symmetry.

Despite the above successes, the electron delocalization model has been criticized on the following grounds:

1. The theory neglects the enhancement of the electronic specific heat by correlation between band electrons and by electron–phonon interactions.[361] For example, UAs has an electronic specific heat constant $\gamma \cong 0.0127 \, cal/mole \, K^2$, which is about 6 times higher than that of UP and the theoretical prediction. These effects, if included in the model, would result in a renormalization of the parameters $\mathscr{D}(\epsilon)$, Δ, and G, found for each compound but would not change the fundamental physics.

2. The crystal-field ground state of the $5f^2$ configuration is normally found to be the *nonmagnetic* Γ_1 singlet rather than the Γ_5 triplet assumed in the above model, as D. Lam and R. Troć have pointed out. However, the experiments which determined this were carried out on insulating crystals[222] or organic complexes.[223–225] In the metallic actinides, the signs and magnitudes of the crystal-field parameters are unknown and may be quite different from those

3.2. THEORIES INVOLVING ITINERANT ELECTRONS

of the insulators because of shielding by conduction electrons. For the same reason, the crystal-field parameters may change at an electron localization or delocalization transition. Moreover, two theoretical analyses of neutron diffraction data[72,80] conclude that the uranium ion in US has the $5f^2$ configuration and a *magnetic* ground state, despite the above objection. The magnetism is believed to be induced by a strong internal magnetic exchange field.

3. At valence transitions rare earth compounds typically show large volume discontinuities ($\Delta V/V \cong 10^{-1}$), whereas the moment-jump transition in UP is accompanied by a very small volume change ($\Delta V/V = 2 \times 10^{-4}$).[253] Furthermore, Steinitz and Grunzweig-Genossar[253] point out that the signs of the volume discontinuities in UP at T' and T_N are opposite to what one would predict by analogy to rare earths, where there is usually a good correlation between ionic radius and valence. These objections are, however, based on an analogy between $4f$ and $5f$ electron behavior, an analogy which is highly questionable in view of the much more extended nature of the $5f$ wavefunctions as compared to those of the $4f$ electrons (see Section 3.2.1). In contrast to the rare earths, there appears to be no clear correlation between the actinide valence and the lattice constant in the compounds under consideration.

4. The NMR experiment in UP measures the internal (transferred hyperfine) field at the site of the P nucleus.[1] The internal field will be proportional to the sublattice magnetization if the valence of the U ions remains constant, but the coefficient of proportionality is expected to depend on the valence state as J. Grunzweig-Genossar pointed out.[373] This effect has not been considered in the theory under discussion, because calculations of the transferred hyperfine field are very difficult and unreliable for metallic actinide materials. The magnitude of the effect is thus unknown.

5. There are a large number of adjustable parameters in the theory. This, however, is justified by the fact that there are a large number of interactions of unknown magnitude in these compounds.

Despite these objections, the large number of experimental data quantitatively explained by the electron delocalization theory and the absence of another similarly successful theory indicate that the theory is substantially valid. However, the recent neutron diffraction data[262] on single crystal UAs, revealing a previously unsuspected spin rotation from the ⟨001⟩ direction to the ⟨110⟩ direction on cooling to the moment-jump transition temperature, shows that at least this transition requires further theoretical study.

3.3. A Theory Intermediate to the Localized and Band Models

It has been mentioned in the preceding sections that the single-particle band or virtual bound state models are not appropriate for describing the highly correlated states of $5f$ electrons in many metallic actinide compounds, because the latter states are not the states of single electrons. The intra-atomic Coulomb, spin-orbit, and crystal-field interactions in the metallic actinides are probably as large or larger than the $f\text{-}f$ and $d\text{-}f$ interatomic hopping energies (Eqs. 3.2.1-1 and 2). In such a case, it is reasonable to begin with a fully localized (or crystal-field) description of the f states and then to add the hopping energies leading to some degree of itinerant (or band) motion. In this way, the difficult part of the problem, the intra-atomic correlation, is built in from the beginning.

With the above ideas in mind, we write the Hamiltonian $\mathcal{H}$ of the crystal of N actinide atoms as follows:

$$\mathcal{H} = \sum_{p,i} (\epsilon_p - n_p \zeta) L_{pp}^i + \sum_{p,q,r,s} \sum_{i,j} B_{ij}(pqrs) L_{pq}^i L_{rs}^j, \quad (3.3\text{-}1)$$

where the quantity ϵ_p is the energy of the localized (or crystal-field) state p having n_p electrons localized in the outer s, d, and f shells of a given atom, and ζ is the Fermi level. The symbol L_{pq}^i denotes a standard basis[374] or atomic[375] operator causing a transition of the atom at site i from the localized state q to the state p. Note that the states p and q may or may not differ in the number of localized electrons. The first term in $\mathcal{H}$ [Eq. (3.3-1)] describes an ensemble of atoms which do not explicitly interact but which weakly exchange both electrons and energy in coming to thermal equilibrium. If only the first sum in $\mathcal{H}$ is present, i.e., if $B_{ij}(pqrs) = 0$, the occupation probability D_p of the atomic state p is given by the grand canonical distribution (Z = Partition function)

$$D_p = Z^{-1} \exp(-(\epsilon_p - n_p \zeta)/k_B T), \quad (3.3\text{-}2)$$

which yields the statistics of the crystal-field model. The second sum in $\mathcal{H}$ contains the interatomic electronic hopping. For example, the effect of the overlap between the f orbitals of different atoms is as follows. An atom at site j in the state s of (say) $5f^n$ may "lose" one electron and change to the state r of $5f^{n-1}$. The emitted electron hops to site i, changing the ion there from the state q of (say) $5f^n$ to the state p of $5f^{n+1}$. There will be an eigenstate of the solid in which the above excitation propagates in the lattice with a certain wavevector and energy. This excitation is analogous to an itinerant Bloch wave in the customary band model. The coefficient (with dimensions

3.3. THEORY INTERMEDIATE TO LOCALIZED AND BAND MODELS

of energy) for the hopping process is called $B_{ij}(pqrs)$, which is also proportional to the width of the corresponding band of excitations.

Each of the symbols p, q, r, and s represents a different set of quantum numbers L, S, J, and Γ_α (Γ_α labels the crystal-field symmetry) and occupation numbers of the $5f$, $6d$, and $7s$ subshells. The states p are linear combinations of determinants of one-electron atomic orbitals and are determined by the procedures of atomic physics.[376] The effects of correlations on the electronic energy bands in the present model appear through the dependence of the hopping energies $B_{ij}(pqrs)$ on the quantum numbers of the atomic states p, q, r and s. Because of selection rules, a number of the $B_{ij}(pqrs)$ may be zero in which case the corresponding excitations will not propagate in the lattice. These excitations remain as narrow atomic levels ("resonances") in the electronic spectrum,[377] and their presence is one way in which the model explains the mixed itinerant and localized characteristic of the $5f$ electrons. An explicit expression for the $B_{ij}(pqrs)$ in the case where only one-electron hopping is considered is the following:

$$B_{ij}(pqrs) = \sum_{m,m'} \sum_{l,l',\sigma} T_{ij}(lm, l'm') \langle ip|c_i^+(l'm'\sigma)|iq\rangle \langle jr|c_j(lm\sigma)|js\rangle. \quad (3.3\text{-}3)$$

where the operator $c_j(lm\sigma)$ is the customary annihilation operator of a *single* electron at site j in the atomic orbital with quantum numbers l, m, and σ. The symbol $|js\rangle$ represents the *many-electron* wavefunction of the localized atomic state s at site j. The quantity $T_{ij}(lm, l'm')$ is the hopping energy between two one-electron atomic orbitals on sites i and j and has the same meaning in the well-known tight-binding band model.

The effects of direct overlap between f orbitals result from those terms in the sum for which $l = l' = 3$. Similarly, the $d\text{-}f$ hybridization is obtained from the $l = 2$, $l' = 3$ terms, etc. The matrix elements in Eq. (3.3-3) may be evaluated by the method of fractional parentage.[376]

At first sight, practical application of the above model may seem hopeless, because there are a large number of localized states p, even for a given configuration, say, $5f^3$. However, it is reasonable in the case of strong intra-atomic and crystal-field interactions to consider only the localized states of lowest energy. For example, for a narrow f band in the actinide intermetallics, it is likely that only three configurations or "polar" states (e.g., $5f^2$, $5f^3$ and $5f^4$) will be important. The simplest approximation considers only the four crystal-field states shown in Fig. 119. Furthermore, the energy $T_{ij}(lm, l'm')$ may be specified by a few variable parameters, as Slater and Koster have shown.[378] The quantities $B_{ij}(pqrs)$ are then completely determined by Eq. (3.3-3).

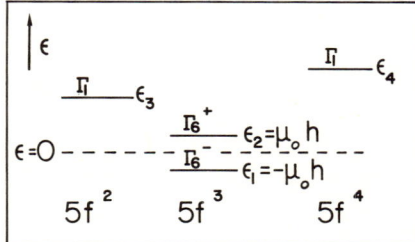

Fig. 119. The crystal-field levels assumed in Ref. 391 to have the lowest energy for each of three configurations of the $5f$ atomic shell. The symbols Γ_1 and $\Gamma_6^\pm$ denote, respectively, a singlet and the components of a doublet having magnetic moment μ_0. h = applied magnetic field.

The model described above is essentially the same as the "atomic representation" of Hubbard[375] and Van Vleck[379] and is similar to the "configuration-based approach" of L. L. Hirst[380] and the old "polarity fluctuation" model.[381] The latter three models have suffered from the drawback that the calculations of the excitation spectrum and the thermodynamic observables of the solid are very difficult. However, recent work[374,377] has provided a systematic Green's function method for treating the Hamiltonian, Eq. (3.3-1). The latter work concerns only Boson excitations (spin waves), but is easily generalized to the case of Fermion excitations (energy bands). According to this method, the equations of motion for the Green's functions of the standard-basis operators L_{pq}^i are written down and solved in the RPA approximation. The poles of the Fourier-transformed Green's functions yield the electronic spectrum, from which the physical observables are calculated. An important result of this procedure is that it interpolates between the Boltzmann statistics of the localized (crystal-field) model and the Fermi–Dirac statistics of the band model.[382,383] The difference between the latter two statistics has hampered the development of the previous models.[380]

As a simple example of the theory under discussion, we consider a model[382] of a simple cubic lattice of actinide ions, each having two exterior s (or d) orbitals and two f orbitals. The model includes nearest neighbor s-s electronic hopping, leading to a broad s band, and s-f hopping, leading to s-f hybridization. The intra-atomic repulsion between two f electrons and between an s and an f electron on the same site is also considered. The inverse paramagnetic susceptibility χ^{-1} of the model is shown in Fig. 120 as a function of temperature T for values of the model parameters appropriate for intermetallic actinides such as UAl_4. Here W is an arbitrary unit of energy of the order of the s band width, d is the s-f admixture probability, and the inverse susceptibility is measured in units of W/μ^2. Note the transition from a nearly temperature-independent susceptibility at low T to a Curie–Weiss behavior as T increases, behavior typical of actinide intermetallics (see Figs. 62, USn_3, and 65). The

3.3. THEORY INTERMEDIATE TO LOCALIZED AND BAND MODELS

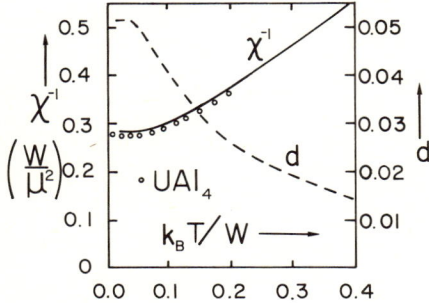

Fig. 120. Circles: Experimental inverse magnetic susceptibility χ^{-1} per U atom of UAl$_4$ vs. T, from Ref. 144. $\mu = 1.63\,\mu_B$; $W = 1333\,k_B$K. Solid curve: theoretical calculation of χ^{-1} for the narrow-band model described in 3.3, Ref. 382. Dashed curve: the coherent hybridization parameter d vs. T calculated by the above model, showing the "thermal dehybridization" effect.

physical reason for the transition is that at low T the f states are hybridized with the s states, yielding narrow bands intersecting the Fermi level. As T increases, the hybridization, i.e., the coherent admixture of the s and f states, decreases rapidly as shown in Fig. 120 (dashed curve). The f electrons become more localized and therefore $\chi^{-1}(T)$ develops the Curie–Weiss behavior characteristic of localized electrons.

Thus the model yields a "thermal dehybridization" effect very similar to that predicted by Doniach's theory.[365] A significant difference is that Doniach uses single-electron operators and obtains the dehybridization and electrical resistivity vs. T behavior by incorporating in the electronic energies an imaginary part having a T^2 temperature dependence. By contrast, the present model uses the standard-basis operators, and the electronic energies are real.

The electrical resistivity ρ of the present model has been estimated under the crude assumption that only electrons in s orbitals contribute to the conductivity and that the localized f electrons act as scattering centers.[382] The resulting temperature dependence of ρ shows qualitative agreement with experiments on a large number of actinide intermetallics previously described as "spin-fluctuation" materials (see 3.2.2 for a discussion of the spin-fluctuation concept).

The above results apply to those intermetallics that do not order magnetically. A preliminary calculation for the magnetically ordered actinide intermetallics such as UPt[157] and NpOs$_2$[155] has been carried out taking into account the spin-orbit and crystal-field interactions.[383] The results indicate that the magnetic moments, susceptibilities, and isomer shifts observed in these materials[155] (see 2.5) are also explicable with the present model.

Razafimandimby and Erdös[443] (RE) consider the effects of a finite overlap energy τ_{lm} between f orbitals on sites l and m in a mixed-valence rare earth or actinide solid. The band calculations of Pickett et al.[444] for Ce metal, in which

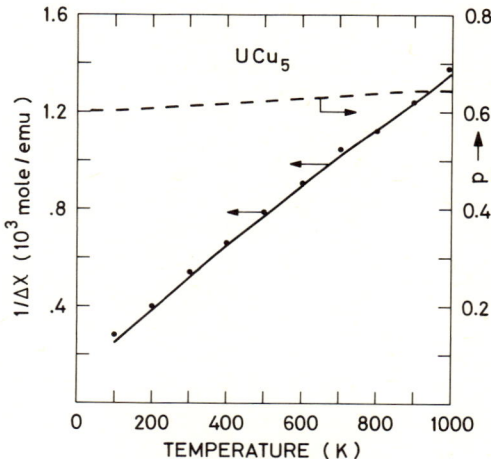

Fig. 121. Left ordinate: Inverse of the paramagnetic susceptibility difference $\Delta\chi = \chi - 0.77 \times 10^{-3}$ emu/mole as a function of temperature for UCu_5. Full dots: Measurement of χ (from Ref. 299). Full line: Calculated from the theoretical model with parameters as given in Ref. 443. Right ordinate and broken line: Probability p of f^n configuration of the U ion, calculated with the same values of the parameters.

the f bandwidth was found to be $\cong 1$ eV, show that the usually neglected f-f overlap may be significant. The RE Hamiltonian $\mathcal{H}$ is written as follows:

$$\mathcal{H} = \sum_{a=0}^{r} \epsilon_a L_{aa}^l + \sum_{k\sigma} \epsilon_{k\sigma} c_{k\sigma}^+ c_{k\sigma} + \sum_{l\sigma} \sum_{kk'} G_a e^{i(k-k')\cdot R_l} c_{k\sigma}^+ c_{k'\sigma} L_{aa}^l \quad (3.3\text{-}4)$$

Here $c_{k\sigma}^+$ ($c_{k\sigma}$) is a creation (annihilation) operator for an electron in a band state of energy $\epsilon_{k\sigma}$; G_a is the short-range repulsion between a band electron and a localized f^n configuration $|a\rangle$ of energy ϵ_a; and t_{lm} is the hopping energy between f orbitals on sites l and m. The localized states $|a\rangle$ are restricted to being a singlet ($a=0$) of the f^{n-1} configuration and an r-tuplet ($a=1,\ldots,r$) of the f^n configuration. This set of states describes Ce^{3+} ions where $n=1$ and $r=2j+1=6$, and U^{3+}, where $n=3$ and $r=6$ (assuming a degenerate Γ_6 doublet and a $\Gamma_8^{(2)}$ quartet ground-state of $5f^3$).

The random phase approximation is applied to solve for the Green functions of the standard-basis operators, from which the occupation probabilities of the states $|a\rangle$ are calculated. The density of states of the f band is approximated by a Lorentzian of half-width t. With reasonable values of the param-

3.3. THEORY INTERMEDIATE TO LOCALIZED AND BAND MODELS

eters the calculations achieve good agreement with the experimental paramagnetic susceptibility and valence (as deduced from thermal expansion and photoemission data) of $CePd_3$, UCu_5, and $UNi_{0.4}Cu_{4.6}$ as functions of temperature (see Fig. 121). The success of these calculations provides further evidence for the presence of mixed valence in the Ce and uranium compounds.

In summary, a theory based on Hubbard's atomic representation and the standard-basis operators seems to be the most promising approach to interpreting the magnetic and electronic properties of actinide materials. The experience indicates that the traditional single-particle band and virtual bound states models have limited applicability to actinides because of the highly correlated, many-particle nature of the states of the $5f$ electrons.

3.4. Similarities between Mixed-Valence Rare Earths and Metallic Actinide Compounds

The same model as the one described in Section 3.3 has been applied to mixed-valence rare earth intermetallics such as YbC_2,[392] in which a $4f$ electron is thermally promoted to the $5d$-$6s$ conduction band. The magnetic susceptibilities and electrical resistivities of the latter compounds show temperature dependencies which resemble those observed in "spin-fluctuation" actinide intermetallics (see Figs. 61, 62, 69, 121). There are other striking similarities between metallic actinides and mixed-valence rare earths. For example, in both classes of materials, the crystal-field coefficients are often found to have a sign opposite to that which would be expected from a point-charge model (example: TmSe, USb, and PuP). The electronic specific heat in the mixed-valent phase is very large in rare earths and actinides (see Table 3). Valence changes may be induced by alloying (e.g., $SmY_{1-x}S_x$,[397] $UNi_{5-x}Cu_x$[173]) temperature (e.g., Ce metal,[393] UP[16]), or pressure (e.g., SmS,[394] possibly UPt[157]). In the mixed-valence state, the magnetic moment is often anomalously small (e.g., TmSe,[395] $AmFe_2$[159]). Finally, photoemission data show the presence of narrow f levels at or very near the Fermi level, and in the case of UPS (see p. 13) show structure indicating the presence of two actinide valence states.[436]

A difference between the compounds of rare earths and actinides is that in the former materials there is usually found a close correlation between the lattice constant and the valence state of the rare earth atom. On the other hand, in the actinides, particularly in those that are metallic, the valence may not, in general, be deduced from the lattice constants. In fact, as the actinide–actinide separation decreases, the $5f$ electrons form itinerant bands rather than remain localized in the f shell (see Section 3.1.1). We attribute this difference in behavior to the greater spatial extension of $5f$ electrons as compared to $4f$ electrons. The $4f$ shell is more shielded from the external environment, and thus for rare earth atoms the concept of an ionic radius is more valid.

The striking similarities between the mixed-valence rare earths and the metallic actinide compounds is explored in greater detail in a 1979 review article.[385] These similarities indicate that our understanding would be enhanced by comparative studies putting the two classes of materials on an equal footing. For example, there are many studies of the effects of pressure in inducing valence transitions in rare earths, but similar studies on actinides are rare. So far pressure studies have only been made on US,[384] UN,[263] UAs,[262] UPt,[157] and $NpOs_2$.[280]

APPENDIX A

Irreducible Tensor Operators

Generally, irreducible tensor operators $\mathbf{T}_q^k$ of rank k with $2k+1$ ($q = -k, \ldots, k$) components are defined as 1) being quantum mechanical operators; and 2) obeying the commutation rules with the total orbital angular moment operator $\mathbf{L}$

$$[\mathbf{L}^{\pm}, \mathbf{T}_q^k] = \sqrt{(k \mp q)(k \pm q + 1)}\, \mathbf{T}_{q \pm 1}^k, \tag{A-1}$$

$$[\mathbf{L}^z, \mathbf{T}_q^k] = q\, \mathbf{T}_q^k. \tag{A-2}$$

It follows that the $2k+1$ components transform among themselves according to the $2k+1$ dimensional representation of the rotation group; furthermore, the components $\mathbf{T}_q^k$ are hermitean operators, and the Wigner–Eckart theorem holds. Irreducible spherical tensors are irreducible tensor operators constructed out of spherical harmonics.

Racah[386] introduced unit tensor operators which we will define as follows:

$\mathbf{u}_q^k$ ($q = -k, \ldots, k$) is a one-electron unit tensor operator if it is an irreducible tensor operator of rank k (i.e., it obeys the commutation relations A-1,2, with $l^{\pm}$ and l^z), and as an operator acting on single-electron wavefunctions $|\alpha l m\rangle$ (l = orbital quantum number, m = azimuthal quantum number, α = collection of all other quantum numbers) it is diagonal and normalized with respect to α and l. Hence, by virtue of the Wigner–Eckart theorem[329,331]

$$(\alpha l m | \mathbf{u}_q^k | \alpha' l' m') = (-1)^{l-m} \begin{pmatrix} l & k & l \\ -m & q & m' \end{pmatrix} \delta_{\alpha\alpha'} \delta_{ll'} \tag{A-3}$$

where the bracket with six entries is the Wigner $3j$ symbol.[329,331]

We note that these matrix elements exist only if $k \leq 2l$; and $-m + q + m' = 0$.

Unit tensor operators acting on n-electron states are defined as

$$\mathbf{U}_q^k = \sum_{i=1}^n \mathbf{u}_q^k(i) \tag{A-4}$$

where $\mathbf{u}_q^k(i)$ is the unit tensor operator as defined above, acting on the ith electron.

We may represent the one-particle spin operators also as irreducible tensor operators. It suffices to notice that if we identify s^+, s^-, and s^z with $\mathbf{T}_1^1$, $\mathbf{T}_{-1}^1$, and $\mathbf{T}_0^1$, respectively, and replace $\mathbf{L}$ also by $\mathbf{s}$, the commutation relations Eqs. A-1,2 coincide for any value of s with those for the spin angular momentum operators. Therefore, to conform with the tensor notation we relabel $\mathbf{s}^+$, $\mathbf{s}^-$, and $\mathbf{s}^z$ as $\mathbf{s}_1^1$, $\mathbf{s}_{-1}^1$, and $\mathbf{s}_0^1$, and define

$$\mathbf{s}_p^1(i) \quad p = -1, 0, 1$$

as the irreducible one-particle spin operator for particle i, and

$$\mathbf{S}_p^1 = \sum_{i=1}^n \mathbf{s}_p^1(i) \tag{A-5}$$

as the n-particle irreducible spin-operator. Spin and orbital tensors commute.

The advantages of unit tensor operators are that 1) all relevant operators which occur in the Hamiltonian may be expressed in terms of them; 2) their matrix elements may be constructed as a product of a Clebsch–Gordan coefficient[321] $\begin{pmatrix} L & k & L' \\ M & q & M' \end{pmatrix}$ and a reduced matrix element $(l^n \alpha LSJ \| \mathbf{U}_q^k \| l^n \alpha' L'S'J')$; and 3) the reduced matrix elements have been tabulated for all possible configurations of p, d, and f electrons.[318] Hence the construction of the Hamiltonian matrix for an ion is reduced to the reading of tables.

One example of the application of irreducible tensor operators was given in Section 3.1.2c, where the spin–orbit coupling was expressed as

$$\mathcal{H}_{so} = \gamma \sum_{i=1}^n \sum_{q=-1}^1 s_q^1(i) u_{-q}^1(i) \equiv \gamma \sum_{q=-1}^1 (-1)^q \mathbf{V}_{q,-q}^{11}. \tag{A-6}$$

It is easy to show that $\mathbf{V}_{pq}^{11}$ is an irreducible tensor operator because it obeys the commutation relations, Eqs. A-1 and 2 with $\mathbf{L}$ and also with $\mathbf{S}$. The matrix elements of this operator are given by Eq. 3.1.2-11, and the reduced matrix elements are tabulated.[318]

The crystal-field Hamiltonian $\mathcal{H}_{cf}$ can also be expressed in terms of unit tensor operators. For this purpose $\mathcal{H}_{cf}$ is first written as a sum of n-electron irreducible tensor operators defined by Racah as

$$C_q^k \equiv \sqrt{\frac{4\pi}{2k+1}} \sum_{i=1}^{n} Y_k^q(i). \qquad (A\text{-}7)$$

Their matrix elements are

$$(l^n \alpha LSJM|C_q^k|l^n \alpha' L'S'J'M') = (2l+1)(-1)^{k-l}\begin{pmatrix} l & l & k \\ 0 & 0 & 0 \end{pmatrix}(LSJM|U_q^k|L'S'J'M') \qquad (A\text{-}8)$$

and, following Edmonds[329]

$$(l^n \alpha LSJM|U_q^n|l^n \alpha' L'S'J'M') = \delta_{ss'}(-1)^{k+S+L'+2J-M}\sqrt{(2J+1)(2J'+1)}$$

$$\times \begin{pmatrix} J & k & J' \\ -M & q & M' \end{pmatrix} \begin{Bmatrix} L & J & S \\ J' & L' & k \end{Bmatrix} (l^n \alpha SL\|U^k\|l^n \alpha' SL'). \qquad (A\text{-}9)$$

Note that in addition to the factoring out of the Wigner-6j-symbol (round bracket) by virtue of the Wigner–Eckart theorem, it was possible to factor out the dependence on J and J' in the form of the Racah coefficient (curly bracket). The remaining reduced matrix element depends only on l, n, L, S, and the quantum numbers symbolized by α, whose significance is explained in the main text; and is tabulated by Nielson and Koster.[318]

We remark that spherical harmonics acting between wavefunctions belonging to different configurations, i.e., l^n and $l^{n-1}l'^1$, cannot be expressed by unit tensor operators since the latter are diagonal in l, l'.

APPENDIX B

Magnetic Structures

General

The results of elastic neutron diffraction, magnetization, magneto-optical and other experiments on actinide compounds can most often be interpreted in terms of an ordered magnetic structure that consists of magnetic moments localized at the actinide ions and, in certain cases, on other ions as well. An example of the latter case is $UFeO_4$, where both U and Fe have moments.

Ferromagnetic (Fig. 4a) and several types of antiferromagnetic (Fig. 4b, c) structures have been observed. Further, there exist ferrimagnetic compounds in which the moments on different sublattices do not compensate each other. Spiral magnetic structures, such as observed in several rare earth metals, have not been seen in actinide compounds. On the other hand, new types of structures, first proposed in another context by Kouvel and Kasper[104] and called 2k and 3k structures have been observed in actinide compounds. In addition, structures that are incommensurate with the crystal structure have also been reported. The application of a magnetic field or uniaxial pressure or substitutional alloying can cause transformations from one of these structures to another. We will discuss some general aspects of these structures.

An *incommensurate* structure is one in which the localized magnetic moments are arranged in a spatially repetitive pattern, but a repetition whose period is not an integral multiple of the lattice period. A formal definition of commensurability will be given in Eq. (A-3) below.

An important distinction exists between structures where the ionic magnetic moments have a fixed magnitude, and structures where the magnitude (modulus) of the moment is modulated (e.g., sinusoidally).

The first case is most often observed at low T because crystallographically equivalent ions, being in states with a definite configuration and J value, will have the same modulus m of their saturation moment. In order to form an incommensurate structure, such moments have to make a *variable angle* with some crystal axis. It is this angle which varies with an incommensurate period along the lattice. Hence, there cannot exist collinear (i.e., constant angle) incommensurate magnetic structures in the case of moments of fixed magnitude. If the moments of fixed magnitude lie in one plane, fan-type structures result; if they are not restricted to a plane spirals or more general arrangements arise.

The situation is different if the moments on crystallographically equivalent ions may have different magnitudes. This case may occur if the crystal-field or exchange interaction mixes different J values, or if the ion is in a mixed configuration, or the exchange field varies periodically along the lattice, or if conduction electron polarization contributes to the local moment. In this case, besides the above-mentioned structures (fan, spiral, etc.), a *collinear* incommensurate magnetic structure may also arise — the magnitude of the moment is modulated. USb_xTe_{1-x} among the actinides, and the rare earths Er and Tm may be examples of this kind.

The following discussion will be restricted to the case of ionic moments of fixed magnitude.

Fourier Analysis of Magnetic Structures

Let us assume that the magnetic structure is periodic. If the period *is commensurate* with the lattice, the magnetic moment density $\mathbf{m}(\mathbf{x})$ inside the crystal has the property

$$\mathbf{m}(\mathbf{x} + \mathbf{c}) = \mathbf{m}(\mathbf{x}), \tag{B-1}$$

with

$$\mathbf{c} = n_1 \mathbf{c}_1 + n_2 \mathbf{c}_2 + n_3 \mathbf{c}_3, \tag{B-2}$$

where n_i are integers, and $\mathbf{c}_i$ are the primitive lattice vectors of the magnetic cell. The latter vectors have the property of being the three shortest possible linearly independent vectors for which Eq. (B-1) holds. We consider the practically important case[†] when it is possible to express the primitive magnetic lattice vectors as integral multiples of the primitive crystallographic lattice vectors $\mathbf{a}_i$, i.e.,

$$\mathbf{c}_i = n_i \mathbf{a}_i \quad i = 1, 2, 3; \quad \text{where } n_i = \text{positive integers.} \tag{B-3}$$

In an incommensurate magnetic structure the n_i are *not* integers. The volume v_m of the magnetic cell in terms of the volume v of the primitive crystal cell is

$$v_m = n_1 n_2 n_3 v. \tag{B-4}$$

[†] In the general case, when the primitive magnetic lattice vectors are linear combinations with integral coefficients of $\mathbf{a}_i$, the calculations are more involved, but the conclusions remain the same.

MAGNETIC STRUCTURES

Let us suppose that the ions are localized at the sites R_i. We shall now replace the site index i by two new indices. The first of these, a capital letter, will number the magnetic cell in which the ion is located, and the second index, a small greek letter, numbers the magnetic ion in the cell. If there are s localized moments per magnetic cell, we have

and
$$i \to J, \nu; \quad i = 1, \ldots, N; \quad \nu = 1, \ldots, s; \quad J = 1, \ldots, N/s$$
$$s = n_1 n_2 n_3. \tag{B-5}$$

If the magnetic moments are localized, the knowledge of the moments $m(R_i) = m(R_{J\nu})$ for $\nu = 1, \ldots, s$ within one magnetic unit cell completely specifies the magnetic structure.

Neutron diffraction experiments do not yield any direct information concerning the moments $m(x_i)$, but give the squared modulus of their Fourier components. It is for this reason only that one needs to describe magnetic structures using Fourier analysis.

The periodicity of $m(x)$ expressed by Eq. (B-1) permits its development into a Fourier series of the form

$$m(x) = v_m^{-1} \sum_K m(K) e^{iK \cdot x} \tag{B-6}$$

The sum is extended over all vectors K of the magnetic reciprocal lattice. These vectors are of the form

and
$$K = m_1 K_1 + m_2 K_2 + m_3 K_3, \quad m_i = \text{integers}, \tag{B-7}$$
$$K_r = 2\pi v_m^{-1} c_s \times c_t. \tag{B-8}$$

The subscripts r, s, and t represent any cyclic permutation of the indices 1, 2, and 3. Since the magnetic moment is real $m^*(R_{I\alpha}) = m(R_{I\alpha})$, and it follows

$$m^*(-K) = m(K). \tag{B-9}$$

The factor v_m^{-1} in Eq. (B-6) is introduced to simplify later formulas.

A great simplification of Eq. (B-6) arises due to the commensurability expressed by Eq. (B-3) of the magnetic and crystallographic lattices. It follows from Eqs. (B-3), (B-4), and (B-8) that the vectors K_r are of the form

$$K_r = \frac{1}{n_r} k_r \tag{B-10}$$

where k_r are the crystallographic primitive reciprocal lattice vectors defined by

$$k_r = 2\pi v^{-1} a_s \times a_t \tag{B-11}$$

with r, s, t being cyclic permutations of 1, 2, and 3. Hence, the primitive vectors of the magnetic reciprocal lattice are fractions of the primitive vectors of the crystallographic reciprocal lattice.

It follows that if we write the integers m_i which occur in Eq. (B-7) as integral multiples of n_i plus a remainder

$$m_i = p_i n_i + q_i, \quad 0 \leq q_i < n_i \quad (i = 1, 2, 3) \tag{B-12}$$

then every vector **K** of the magnetic reciprocal lattice may be decomposed as

$$\mathbf{K} = \sum_{i=1}^{3} p_i \mathbf{k}_i + \sum_{i=1}^{3} \frac{q_i}{n_i} \mathbf{k}_i, \quad q_i < n_i . \tag{B-13}$$

The first term in Eq. (B-13) is a vector **k** of the crystallographic reciprocal lattice, whereas the second term is a vector **κ** which, if placed at the origin, will lie entirely inside the unit cell of the crystallographic reciprocal lattice.

Hence, we write

$$\mathbf{K} = \mathbf{k} + \boldsymbol{\kappa}. \tag{B-14}$$

This decomposition facilitates the simplification of the Fourier transform, Eq. (B-16), and allows us to find its inverse.

We multiply Eq. (B-6) by $e^{-i\mathbf{K}' \cdot \mathbf{x}}$ and integrate over the volume of one magnetic cell v_m. Since

$$\int_{v_m} \exp i (\mathbf{K} - \mathbf{K}') \cdot \mathbf{x} \, d^3 x = v_m \Delta(\mathbf{K} - \mathbf{K}') \tag{B-15}$$

it follows that

$$\mathbf{m}(\mathbf{K}) = \int_{v_m} \mathbf{m}(\mathbf{x}) e^{-i\mathbf{K} \cdot \mathbf{x}} d^3 x. \tag{B-16}$$

Here, $\Delta(\mathbf{K}) = 1$ if $\mathbf{K} = 0$, and $\Delta(\mathbf{K}) = 0$ if $\mathbf{K} \neq 0$.

In general, an infinity of different vectors $\mathbf{m}(\mathbf{K})$ for all vectors **K** of the magnetic reciprocal lattice have to be known in order to determine $\mathbf{m}(\mathbf{x})$ through Eq. (B-6). However, if the magnetic moments may be considered as localized at the lattice points, matters simplify. Localization means

$$\mathbf{m}(\mathbf{x}) = \sum_{I,\alpha} \mathbf{m}(\mathbf{R}_{I\alpha}) \delta(\mathbf{x} - \mathbf{R}_{I\alpha}) \tag{B-17}$$

If in Eq. (B-16) we choose the cell number $I = 0$, for the localized case we find

$$\mathbf{m}(\mathbf{K}) = \sum_{\alpha} \mathbf{m}(\mathbf{R}_{0\alpha}) e^{-i\mathbf{K} \cdot \mathbf{R}_{0\alpha}} . \tag{B-18}$$

The Fourier transform $\mathbf{m}(\mathbf{K})$ is in the localized case a periodic function —

its period is the unit cell of the crystallographic (not magnetic!) reciprocal lattice. In fact,

$$m(K + k) = \frac{1}{v_m} \sum_\alpha m(R_{0\alpha}) e^{-i(K+k) \cdot R_{0\alpha}} = m(K), \qquad (B-19)$$

since $\exp(i k \cdot R_{0\alpha}) = 1$.

As a consequence, it is sufficient to know $m(\kappa)$, where κ is of the form

$$\kappa = \sum_{l=1}^{3} q_e K_l, \quad 0 \leq q_l < n_l \qquad (B-20)$$

The number of vectors κ is equal to the number s of (in general different) magnetic moments per magnetic unit cell. Hence, if by neutron diffraction one can determine the s vectors $m(\kappa)$ for the s wavevectors κ which lie in the interior of the unit crystallographic reciprocal cell, all the localized magnetic moments can be found from

$$m(R_{0\alpha}) = \sum_\kappa m(\kappa) e^{i\kappa \cdot R_{0\alpha}}. \qquad (B-21)$$

For completeness, we rewrite the inverse formula

$$m(\kappa) = s^{-1} \sum_\alpha m(R_{0\alpha}) e^{-i\kappa \cdot R_{0\alpha}}. \qquad (B-22)$$

Eq. (21) is obtained readily by integrating Eq. (B-17) over an infinitesimal volume enclosing $R_{0\alpha}$, and making use of Eqs. (B-6), (B-15), (B-17), and (B-19).

We note that by addition of vectors k of the crystallographic reciprocal lattice the set of s vectors κ may be substituted by an equivalent set κ', which will lie in or on the boundary of the first Brillouin zone of the crystallographic lattice. For the vectors κ' Eq. (B-20) is no more valid, since negative coefficients q_l will also occur.

Elastic Neutron Diffraction

The scattering cross-section of neutrons exhibits peaks when the scattering vectors, i.e., the change in the neutron wavevector, is equal to a vector of the reciprocal lattice, and when certain other conditions are met (see below). The scattering of the neutron by the nuclear forces gives rise to Bragg peaks at wavevectors k of the crystallographic reciprocal lattice, whereas scattering by the magnetic moments of the ions produces Bragg peaks at wavevectors **K** of the magnetic reciprocal lattice.

Theory shows[451] that if all magnetic ions are of the same kind, the magnetic scattering intensity I at a magnetic reciprocal lattice vector **K** is proportional to

$$I(\mathbf{K}) \propto |F(\mathbf{K})\,\mathbf{m}(\mathbf{K})_\perp|^2. \qquad \text{(B-23)}$$

Here $F(\mathbf{K})$ is the so-called magnetic form factor of the ion in question, and $\mathbf{m}(\mathbf{K})_\perp$ is the component of $\mathbf{m}(\mathbf{K})$ *perpendicular* to the vector **K**. Here $\mathbf{m}(\mathbf{K})$ is the Fourier transform of the magnetic moments as defined by Eq. (B-18). Some general conclusions from Eq. (B-23) concerning neutron diffraction are the following:

    a. Only the component of $\mathbf{m}(\mathbf{K})_\perp$ perpendicular to **K** can be measured; also there is no peak for **K** vectors that are parallel to $\mathbf{m}(\mathbf{K})$.

    b. Since I is proportional to the squared modulus of $\mathbf{m}(\mathbf{K})$, the phase of $\mathbf{m}(\mathbf{K})$ cannot be determined.

    c. The magnetic form factor has to be known in order to find $\mathbf{m}(\mathbf{K})$.

    d. Since $\mathbf{m}(\mathbf{K} + \mathbf{k}) = \mathbf{m}(\mathbf{K})$, when $\mathbf{k}$ = the crystallographic reciprocal lattice vector, it suffices to determine $\mathbf{m}(\kappa)$ for $\kappa = 1, \ldots, s$. However, due to the limitation expressed in (a), one may have to measure $I(\kappa + \mathbf{k})$ for different **k**. For instance, if $\mathbf{m}(\kappa)$ is parallel to κ, $I(\kappa) = 0$, but if a vector **k** is chosen not parallel to κ, then $\mathbf{m}(\kappa + \mathbf{k}) = \mathbf{m}(\kappa)$ is not parallel to $\kappa + \mathbf{k}$, and the measurement of $I(\kappa + \mathbf{k}) \neq 0$ allows the complete determination of $|\mathbf{m}(\kappa)|$.

    e. The determination of the moments at the lattice sites (hence the magnetic order) $\mathbf{m}(\mathbf{R}_{I\alpha})$ from $\mathbf{m}(\mathbf{K})$ is done using the Fourier transform formula of Eq. (B-21). However, one has to bear in mind that ambiguities remain because the phases of the vectors $\mathbf{m}(\mathbf{K})$ are not known.

The scattering of spin-polarized neutrons gives more information useful to determine unambiguously the magnetic structure. For instance, it allows the measurement of the magnetic form factor, or it permits the separation of nuclear and magnetic scattering intensities if they appear at the same scattering vector **K**. For details we refer to the pertinent literature.

Examples

To illustrate the foregoing mathematical analysis we describe two examples. These are only two-dimensional structures to allow short calculations and simple graphical representations.

Example 1

Consider a plane simple square lattice of identical magnetic ions of lattice constant a. The primitive vectors are $a\hat{x}, a\hat{y}$ ($\hat{x}, \hat{y}$ = unit vectors in the directions x, y). The primitive vectors of the crystallographic reciprocal lattice are $\mathbf{k}_1 = (2\pi/a)\hat{x}$, $\mathbf{k}_2 = (2\pi/a)\hat{y}$. Different moments $\mathbf{m}_1, \mathbf{m}_2, \mathbf{m}_3$ are attached to the ions in the periodic arrangement shown in Fig. 122. The magnetic unit cell is obtained by tripling the crystallographic unit cell in the x direction, and in Eq. (B-4)

$$n_1 = 3, \quad n_2 = 1; \quad v_m = 3v.$$

The primitive vectors of the magnetic reciprocal lattice are

$$\mathbf{K}_1 = \tfrac{1}{3}\mathbf{k}_1, \quad \mathbf{K}_2 = \mathbf{k}_2.$$

Hence the integers q_1, q_2 take on the values

$$q_1 = 0, 1, 2; \quad q_2 = 0.$$

Therefore, in the decomposition [Eqs. (B-13) and (B-14)] of every vector of the magnetic reciprocal lattice, there may occur the vectors [see Eq. (B-20)]

$$\boldsymbol{\kappa}_1 = 0, \quad \boldsymbol{\kappa}_2 = \mathbf{K}_1 = \tfrac{1}{3}\mathbf{k}_1, \quad \text{and} \quad \boldsymbol{\kappa}_3 = 2\mathbf{K}_1 = \tfrac{2}{3}\mathbf{k}_1$$

(Instead of $\boldsymbol{\kappa}_3 = \tfrac{2}{3}\mathbf{k}_1$ which lies outside the first Brillouin zone, one may use $\boldsymbol{\kappa}_3' = -\tfrac{1}{3}\mathbf{k}_1$.)

In the Fourier analysis the vectors $\mathbf{m}(\boldsymbol{\kappa}_1)$, $\mathbf{m}(\boldsymbol{\kappa}_2)$, and $\mathbf{m}(\boldsymbol{\kappa}_3)$ are the only independent ones. Using as a basis vectors $\mathbf{R}_{0,\alpha}$ ($\alpha = 1, \ldots, s$) of the magnetic unit cell

$$\mathbf{R}_{0,1} = 0, \quad \mathbf{R}_{0,2} = a\hat{x}, \quad \mathbf{R}_{0,3} = 2a\hat{x}$$

the Fourier vectors are given by Eq. (B-22) as

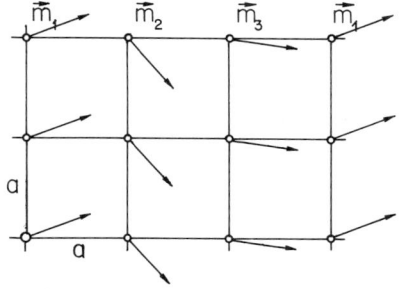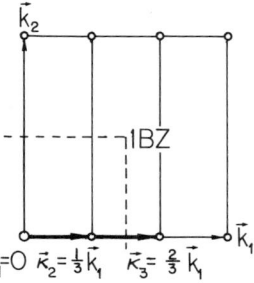

Fig. 122. Illustration of a planar magnetic structure with tripling of the crystallographic unit cell in the x direction. Three arbitrary magnetic moments $\mathbf{m}_1, \mathbf{m}_2, \mathbf{m}_3$ form a repetitive pattern. On the right, the vectors of the reciprocal lattice are shown as defined in the text.

$$m(\kappa_1 = 0) = \tfrac{1}{3}(m_1 + m_2 + m_3) \quad \text{"ferromagnetic component"},$$
$$m(\tfrac{1}{3}k_1) = \tfrac{1}{3}(m_1 + e^{-2\pi i/3} m_2 + e^{2\pi i/3} m_3),$$
$$m(\tfrac{2}{3}k_1) = \tfrac{1}{3}(m_1 + e^{2\pi i/3} m_2 + e^{-2\pi i/3} m_3).$$

With the inversion formula of Eq. (B-21) one recovers $m(R_i)$. This structure would be called a "$1k$ structure," even though three vectors ($\kappa_1, \kappa_2, \kappa_3$) of the reciprocal lattice determine it, but the three vectors are *collinear*.

Neutron diffraction would yield the squared moduli $|m(K_1 = 0)|^2$, $|m(\tfrac{1}{3}k_1)|^2$, and $|m(\tfrac{2}{3}k_1)|^2$. It is left for the reader to analyze the different moment arrangements $\{m_1, m_2, m_3\}$ which are compatible with the same measured set of squared moduli, besides the one shown in the figure. If $m_1^2 = m_2^2 = m_3^2$, i.e., if the moments have equal magnitude, the analysis is considerably simplified.

Example 2

Four different moments $m_1, \ldots, m_4$ are attached to the corners of the unit cell of the square lattice of Example 1, and this arrangement is periodically extended. Here
$$n_1 = 2, \quad n_2 = 2, \quad v_m = 4v.$$
One finds
$$\kappa_1 = 0, \quad \kappa_2 = \tfrac{1}{2}k_1 = \frac{\pi}{a}\hat{x}, \quad \kappa_3 = \tfrac{1}{2}k_2 = \frac{\pi}{a}\hat{y},$$
$$\kappa_4 = \tfrac{1}{2}(k_1 + k_2) = \frac{\pi}{a}(\hat{x} + \hat{y}).$$

All four vectors κ_i lie in or on the boundary of the Brillouin zone. Eq. (B-22) yields

$$m(\kappa_1) = m(0) = \tfrac{1}{4}(m_1 + m_2 + m_3 + m_4),$$
$$m(\kappa_2) = m\left(\frac{\pi}{a}\hat{x}\right) = \tfrac{1}{4}(m_1 - m_2 + m_3 - m_4),$$
$$m(\kappa_3) = m\left(\frac{\pi}{a}\hat{y}\right) = \tfrac{1}{4}(m_1 + m_2 - m_3 - m_4),$$
$$m(\kappa_4) = m\left(\frac{\pi}{a}\hat{x} + \frac{\pi}{a}\hat{y}\right) = \tfrac{1}{4}(m_1 - m_2 - m_3 + m_4).$$

We note two special cases of Example 2 with analogues in the actinide compounds:

1. All moments have equal magnitude, $|m_i| = $ const.; no magnetic reflection is observed for **K** parallel to $\hat{x}$ or to $\hat{y}$, i.e.,

$$m_x\left(\frac{\pi}{a}\hat{x}\right) = 0, \quad m_y\left(\frac{\pi}{a}\hat{y}\right) = 0, \tag{B-X}$$

and, in addition, $\mathbf{m}(\boldsymbol{\kappa}_4) = 0$. This case is usually called a *2k structure*, since only $\boldsymbol{\kappa}_2$ and $\boldsymbol{\kappa}_3$ are nonzero Fourier components. (The nomenclature does not yet seem sufficiently established to decide whether or not the condition expressed by the last pair of equations should be made part of the definition of a *2k structure*.)

In this case we find that $\mathbf{m}_1$ completely specifies the structure, and the only possibility for $v_m = 4v$ is

$$m_{4x} = m_{2x} = -m_{3x} = -m_{1x} \quad \text{and} \quad m_{4y} = m_{3y} = -m_{2y} = -m_{1y}.$$

2. Same as (1), but without the condition $\mathbf{m}(\boldsymbol{\kappa}_4) = 0$.

It is important to realize, however, that if there is a neutron reflection peak for $\boldsymbol{\kappa} = \pi/a(\hat{x} + \hat{y})$, or equivalent vector, the structure ceases to be a *2k* structure, and the moment arrangement may be much more complicated. One finds, for instance

$$\mathbf{m}_2 = \mathbf{m}(\boldsymbol{\kappa}_1) + m_x(\boldsymbol{\kappa}_2)\hat{x} - m_y(\boldsymbol{\kappa}_3)\hat{y} - \mathbf{m}(\boldsymbol{\kappa}_4).$$

3k structures can be exemplified by obvious generalization of the foregoing to three dimensions.

We remark that the *k*-vectors of observed structures usually belong to the same group-theoretical *star*.

References

1. S. L. Carr, C. Long, W. G. Moulton, and M. Kuznietz, *Phys. Rev. Lett.* **23**, 786 (1969).
2. G. H. Lander, M. H. Mueller, and J. F. Reddy, *Phys. Rev. B* **6**, 1880 (1972).
3. J. Leciejewicz, A. Murasik, and T. Palewski, *Phys. Status Solidi B* **46**, K67 (1971).
4. G. H. Lander, L. Heaton, M. H. Mueller, and K. D. Anderson, *J. Phys. Chem. Solids* **30**, 733 (1969).
5. J. M. Gulick and W. G. Moulton, *Phys. Lett. A* **35**, 429 (1971).
6. R. Troć and Z. Kletowski, *Bull. Acad. Pol. Sci.* (in press).
7. M. B. Brodsky, *A.I.P. Conf. Proc.* **5**, 611 (1972).
8. J. F. Counsell, R. M. Dell, A. R. Junkinson, and J. F. Martin, *Trans. Faraday Soc.* **63**, 72 (1967).
9. J. A. C. Marples, *J. Phys. Chem. Solids* **31**, 2433 (1970).
10. G. H. Lander, M. Kuznietz, and D. E. Cox, *Phys. Rev.* **188**, 963 (1969).
11. B. C. Frazer, G. Shirane, D. E. Cox, and C. E. Olsen, *Phys. Rev.* **140**, A1448 (1965).
12. S. I. Parks and W. G. Moulton, *Phys. Rev.* **173**, 333 (1968).
13. C. Long and Yung-Li Wang, *Phys. Rev. B* **3**, 1656 (1970).
14. S. J. Allen, Jr., *Phys. Rev.* **166**, 530 (1968); **167**, 492 (1968).
15. C. P. Bean and D. S. Rodbell, *Phys. Rev.* **126**, 104 (1962).
16. J. M. Robinson and P. Erdös, *Phys. Rev. B* **8**, 4333 (1973); *Phys. Rev. B* **9**, 2187 (1974). For a short account see also P. Erdös and J. M. Robinson, in: *Magnetism and Magnetic Materials 1972, A.I.P. Conf. Proc. 10*, (C. D. Graham, Jr. and J. J. Rhyne, eds.) (1973), Part 2, p. 1070.
17. J. G. Huber, M. B. Maple, and D. Wohlleben, *A.I.P. Conf. Proc.* **10**, 1075 (1973).
18. A. J. Arko, M. B. Brodsky, and W. J. Nellis, *Phys. Rev. B* **5**, 4564 (1972).
19. J. Grunzweig-Genossar, M. Kuznietz, and F. Friedman, *Phys. Rev.* **173**, 562 (1968).
20. F. Holtzberg, T. R. McGuire, and S. Methfessel, in: Landolt-Boernstein, *Numerical Data and Functional Relationships in Science and Technology* (K. H. Hellwege, ed.), Part III, 4a, p. 110, Springer, New York (1970).
21. W. Trzebiatowski and R. Troć, Preprint no. 43 (75), Institute for Low Temperature and Structure Research, Polish Academy of Sciences, Wroclaw (1971).
22. W. Suski, *J. Phys. (Paris)* **40**, C4-43 (1979).
23. *The Actinides-Electronic Structure and Related Properties* (A. J. Freeman and J. B. Darby, Jr., eds), Academic Press, New York (1974).
24. C. F. van Doorn and P. de V. du Plessis, *J. Magn. Magn. Mat.* **5**, 164 (1977).
25. R. A. Cowley and G. Dolling, *Phys. Rev.* **167**, 464 (1968).

REFERENCES

26. J. C. Spirlet, E. Bednarczyk, I. Ray, and W. Mueller, *J. Phys. (Paris)* **40**, C4-108 (1979).
27. M. Yessik, *J. Appl. Phys.* **40**, 1133 (1969).
28. G. Busch, F. Hulliger, and O. Vogt, *J. Phys. (Paris)* **40**, C4-62 (1979).
29. C. F. Buhrer, *J. Phys. Chem. Solids* **30**, 1276 (1969).
30. C. Keller, *The Chemistry of Transuranium Elements*, Verlag Chemie GmbH, Weinheim, Bergstrasse (1971).
31. D. A. Damien, R. G. Haire, and J. R. Peterson, *J. Phys. (Paris)* **40**, C4-95 (1979).
32. Z. Henkie and J. P. Markowski, *J. Cryst. Growth* **41**, 303 (1977).
33. J. C. Spirlet, *J. Phys. (Paris)* **40**, C4-87 (1979).
34. G. Calestani, J. C. Spirlet, and W. Muller, *J. Phys. (Paris)* **40**, C4-106 (1979).
35. I. A. Izyumov and R. P. Ozerov, *Magnetic Neutron Diffraction*, Plenum Press, New York (1970).
36. C. P. Slichter, *Principles of Magnetic Resonance*, Harper and Row, New York (1963), and 2nd ed. Springer, Berlin (1980).
37. *The Mössbauer Effect* (H. Frauenfelder, ed.), W. A. Benjamin, New York (1962).
38. J. A. Stone, *Np Mössbauer Spectroscopy*, (unpublished report), Savannah River Lab., E. I. du Pont de Nemours and Co, Aiken, S.C. 29801, U.S.A. (1973).
39. A. Schenck, in: *Nuclear and Particle Physics at Intermediate Energies*, (J. B. Warren, ed.), p. 159, Plenum Press, New York (1976).
40. G. H. Lander, B. D. Dunlap, M. H. Mueller, I. Nowik, and J. F. Reddy, *Int. J. Magn.* **4**, 99 (1973).
41. O. L. Kruger and J. B. Moser, *J. Chem. Phys.* **46**, 891 (1967).
42. M. A. Kanter and C. W. Kazmierowicz, *J. Appl. Phys.* **35**, 1053 (1964).
43. C. H. de Novion and R. Lorenzelli, *J. Phys. Chem. Solids* **29**, 1901 (1968).
44. P. Costa and R. Lallement, *Phys. Lett.* **7**, 21 (1963).
45. Z. Fisk and B. R. Coles, *J. Phys. C* **3**, L 104 (1970).
46. J. F. Counsell, R. M. Dell, and J. F. Martin, *Trans. Faraday Soc.* **62**, 1736 (1966).
47. L. F. Bates and P. B. Unstead, *Brit. J. Appl. Phys.* **15**, 543 (1964).
48. O. L. Kruger and J. B. Moser, *J. Phys. Chem. Solids* **28**, 2321 (1967).
49. E. D. Cater, E. G. Rauh, and R. J. Thorn, *J. Chem. Phys.* **35**, 619 (1961).
50. F. A. Wedgwood, C. H. de Novion, and A. P. Murani, *J. Phys. C* **11**, 2723 (1978).
51. M. Kuznietz, G. H. Lander, and Y. Baskin, *J. Appl. Phys.* **40**, 1130 (1969).
52. R. C. Maglić, G. H. Lander, M. H. Mueller, J. Crangle, and G. S. Williams, *Phys. Rev. B* **10**, 1943 (1974).
53. L. Heaton, M. H. Mueller, K. D. Anderson, D. D. Zauberis, *J. Phys. Chem. Solids* **30**, 453 (1969).
54. S. Sidhu, W. Vogelsang, and K. D. Anderson, *J. Phys. Chem. Solids* **27**, 1197 (1966).
55. N. A. Curry, *Proc. Phys. Soc., London* **89**, 427 (1966).
56. M. Kamimoto, Y. Takahashi, and T. Mukaibo, *J. Phys. Chem. Solids* **37**, 719 (1976).
57. W. J. Nellis, A. R. Harvey, G. H. Lander, B. D. Dunlap, M. B. Brodsky, M. H. Mueller, J. F. Reddy, and G. R. Davidson, *Phys. Rev. B* **9**, 1041 (1974).
58. C. J. Schinkel and R. Troć, *J. Magn. Magn. Mat.* **9**, 339 (1978).
59. B. R. Cooper, in: *Solid State Physics* (F. Seitz, D. Turnbull, and H. Ehrenreich, eds.), Vol. XXI, Academic Press, New York (1968).
60. J. Leciejewicz, A. Murasik, and R. Troć, *Phys. Status Solidi B* **30**, 157 (1968).
61. J. Leciejewicz, A. Murasik, T. Palewski, and R. Troć, *Phys. Status Solidi* **38**, K89 (1970).

REFERENCES

62. G. Busch, O. Vogt, and H. Bartholin, *J. Phys. (Paris)* **40**, C4-64 (1979).
63. N. A. Curry, *Proc. Phys. Soc. London* **86**, 1193 (1965).
64. T. Muromura and H. Tagawa, *J. Nucl. Mater.* **79**, 264 (1979).
65. J. O. Scarbrough, H. L. Davis, W. Fulkerson, and J. O. Betterton, Jr., *Phys. Rev.* **176**, 666 (1968).
66. R. Didchenko and F. Gortsema, *Inorg. Chem.* **2**, 1079 (1963).
67. C. F. van Doorn and P. de V. du Plessis, *J. Low. Temp. Phys.* **28**, 391 and 401 (1977).
68. J. A. C. Marples, C. F. Sampson, F. A. Wedgwood, and M. Kuznietz, *J. Phys. C* **8**, 708 (1975).
69. G. H. Lander, M. H. Mueller, D. M. Sparlin, and O. Vogt, *Phys. Rev. B* **14**, 5035 (1976).
70. G. H. Lander, W. G. Stirling, and O. Vogt, *Phys. Rev. Lett.* **42**, 260 (1979).
71. T. M. Holden, E. C. Svensson, W. J. L. Buyers, and G. H. Lander, *J. Phys. (Paris)* **40**, C4-31 (1979).
72. F. A. Wedgwood, *J. Phys. C* **5**, 2427 (1972).
73. F. A. Wedgwood and M. Kuznietz, *J. Phys. C* **5**, 3012 (1972).
74. V. Chechernikov, A. Pechennikov, E. Yarembash, L. Martynova, and V. Slavyanskikh, *Sov. Phys. JETP* **26**, 328 (1968).
75. W. Trzebiatowski and W. Suski, *Bull. Acad. Pol. Sci., Ser. Sci. Chim.* **10**, 399 (1962).
76. W. E. Gardner and T. F. Smith, *Proc. 11th Conf. on Low Temperature Physics*, St. Andrews, (J. F. Allen, D. M. Finalyson, D. M. McCall, eds.), Univ. of St. Andrews, Scotland (1969), p. 1377.
77. J. A. C. Marples, *J. Phys. Chem. Solids* **31**, 2431 (1970).
78. G. H. Lander and M. H. Mueller, *Phys. Rev. B* **10**, 1994 (1974).
79. J. A. C. Marples, private communication.
80. A. J. Freeman, J. P. Desclaux, G. H. Lander, and J. Faber, Jr., *Phys. Rev. B* **13**, 1168 (1976).
81. G. Longworth, F. A. Wedgwood, and M. Kuznietz, *J. Phys. C* **6**, 1652 (1973).
82. E. F. Westrum, Jr., R. R. Watters, H. F. Flotow, and D. W. Osborne, *J. Chem. Phys.* **48**, 155 (1968).
83. In Reference 19, the value of γ for USe should be a factor of ten larger (J. Grunzweig-Genossar, private communication).
84. D. Eastman and M. Kuznietz, *J. Appl. Phys.* **42**, 1396 (1971).
85. W. Eib, M. Erbudak, F. Greuter, and B. Reihl, *Phys. Lett. A* **68**, 391 (1978).
86. W. Eib, M. Erbudak, F. Greuter, and B. Reihl, *J. Phys. C* **12**, 1195 (1979).
87. M. Erbudak, F. Greuter, F. Meier, B. Reihl, and J. Keller, *Solid State Commun.* **30**, 439 (1979).
88. M. Erbudak and J. Keller, *Phys. Rev. Lett.* **42**, 115 (1979).
89. T. A. Sandenaw, R. B. Gibney, and C. E. Holley, Jr., *J. Chem. Thermodyn.* **1973**, 41 (1973).
90. A. T. Aldred, B. D. Dunlap, A. R. Harvey, D. J. Lam, G. H. Lander, and M. H. Mueller, *Phys. Rev. B* **9**, 3766 (1974).
91. G. H. Lander, B. D. Dunlap, D. J. Lam, A. Harvey, I. Nowik, M. H. Mueller, and A. T. Aldred, *A.I.P. Conf. Proc.* **10**, 88 (1973).
92. R. Lallement, P. Costa, and R. Pascard, *J. Phys. Chem. Solid* **26**, 1255 (1965).
93. U. Benedict, C. Dufour, and O. Scholten, *J. Nucl. Mat.* **73**, 208 (1978).
94. D. J. Lam, F. Fradin, and O. Kruger, *Phys. Rev.* **187**, 606 (1969).
95. A. Blaise, J. Fournier, and P. Salmon, C.E.N.G., Grenoble, France (unpublished report).

96. M. Kuznietz, *J. Appl. Phys.* **42**, 1470 (1971).
97. M. Kuznietz and J. Grunzweig-Genossar, *J. Appl. Phys.* **41**, 906 (1970).
98. J. Leciejewicz, A. Murasik, R. Troć, and T. Palewski, *Phys. Status Solidi B* **46**, 391 (1971).
99. M. Tetenbaum, *J. Appl. Phys.* **35**, 2468 (1964).
100. J. F. Counsell *et al.*, Proceedings of a 1967 Symposium on Thermodynamics of Nuclear Materials, International Atomic Energy Agency, Vienna (1968), pp. 385–394.
101. J. Leciejewicz, R. Troć, A. Murasik, and T. Palewski, *Phys. Status Solidi B* **48**, 445 (1971).
102. B. R. Cooper and O. Vogt, *J. Phys. (Paris)* **40**, C4-66 (1979).
103. J. Rossat-Mignod, P. Burlet, O. Vogt, and G. H. Lander, *J. Phys. (Paris)* **40**, C4-70 (1979).
104. J. S. Kouvel and J. S. Kasper, *J. Phys. Chem. Solids* **24**, 529 (1963).
105. A. Murasik, W. Suski, R. Troć, and J. Leciejewicz, *Phys. Status Solidi* **30**, 61 (1968).
106. Z. Henkie and W. Trzebiatowski, *Phys. Status Solidi* **35**, 827 (1969).
107. Z. Henkie and Z. Kletowski, *Acta Phys. Pol. A* **42**, 405 (1972).
108. J. Przystawa, *Phys. Status Solidi* **24**, 313 (1967).
109. J. Leciejewicz, R. Troć, A. Murasik, and A. Zygmunt, *Phys. Status Solidi* **22**, 517 (1967).
110. R. Troć, J. Leciejewicz, and R. Ciszewski, *Phys. Status Solidi* **15**, 515 (1966).
111. R. Ballestracci, E. F. Bertaut, and R. Pauthenet, *J. Phys. Chem. Solids* **24**, 487 (1963).
112. A. Murasik, W. Suski, and J. Leciejewicz, *Phys. Status Solidi* **34**, K157 (1969).
113. J. Przystawa and W. Suski, *Phys. Status Solidi* **20**, 451 (1967).
114. F. Friedman, J. Grunzweig-Genossar, and M. Kuznietz, *Phys. Lett. A* **25**, 690 (1967).
115. H. Adachi and S. Imoto, *J. Phys. Chem. Solids* **34**, 1537 (1973).
116. F. Grønwold, M. R. Zaki, E. F. Westrum, Jr., J. A. Sommers, and D. B. Downie, *J. Inorg. Nucl. Chem.* **40**, 635 (1978).
117. R. Troć and Z. Żołnierek, *J. Phys. (Paris)* **40**, C4-79 (1979).
118. H. Ptasiewicz-Bąk, J. Leciejewicz, and A. Zygmunt, *Phys. Status Solidi A* **47**, 349 (1978).
119. A. Zygmunt and M. Duczmal, *Phys. Status Solidi A* **9**, 659 (1972).
120. C. Bazan and A. Zygmunt, *Phys. Status Solidi A* **12**, 649 (1972).
121. W. Suski, A. Czopnik, and T. Mydlarz, *Phys. Status Solidi A* **12**, 525 (1972).
122. E. F. Westrum, Jr., and F. Grønwold, *J. Inorg. Nuclear Chem.* **32**, 2169 (1970).
123. W. Suski, *Phys. Status Solidi A* **13**, 675 (1972).
124. Y. L. Wang and B. R. Cooper, *Phys. Rev.* **172**, 539 (1968).
125. W. Suski, T. Gibinski, A. Wojakowski, and Z. Czopnik, *Phys. Status Solidi A* **9**, 653 (1972).
126. J. M. Fournier, A. Blaise, and P. Salmon, *Int. Congress on Magnetism*, Mowcow, VI, 65 (1973).
127. Y. Hery, D. A. Damien, M. Haessler, and C. H. de Novion, *Radiochem. Radioanal. Lett.* **32**, 283 (1978).
128. R. Ciszewski, A. Murasik, and R. Troć, *Phys. Status Solidi* **10**, K85 (1965).
129. J. Przystawa, *J. Phys. Chem. Solids* **31**, 2158 (1970).
130. J. Przystawa and E. Pravecki, *J. Phys. Chem. Solids* **33**, 1943 (1972).
131. L. Dobrinski and J. Przystawa, *Phys. Status Solidi* **42**, K15 (1970).

REFERENCES

132. B. Stalinski, Z. Bieganski, and R. Troć, *Phys. Status Solidi* **17**, 837 (1966).
133. R. Troć, J. Mulak, and W. Suski, *Phys. Status Solidi B* **43**, 147 (1971).
134. Z. Henkie and C. Bazan, *Phys. Status Solidi A* **5**, 259 (1971).
135. I. H. Warren and C. E. Price, *Can. Metall. Q.* **3**, 245 (1964).
136. Z. Henkie, *Bull. Acad. Polon. Sci., Ser. Sci. Chim.* **20**, 531 (1972).
137. Z. Henkie and J. Klamut, *Physica* **86-88B**, 991 (1977).
138. W. Trzebiatowski, Z. Henkie, K. P. Bielow, A. S. Dmitrewskij, R. E. Lewityn, and Ju. F. Popow, *Sov. Phys. JETP* **61**, 1522 (1971).
139. V. I. Checkernikov, A. V. Pechennikov, M. E. Barykin, V. K. Slavyanskikh, E. I. Yarembash, and G. V. Ellert, *Sov. Phys. JETP* **25**, 560 (1967).
140. W. Suski, T. Mydlarz, and V. U. S. Rao, *Phys. Status Solidi A* **14**, K157 (1972).
141. D. Raphael and C. H. de Novion, *J. Phys. (Paris)* **30**, 261 (1969).
142. B. D. Dunlap, D. J. Lam, G. M. Kalvius, and J. K. Shenoy, *J. Appl. Phys.* **42**, 1719 (1971).
143. V. I. Kutaiber, N. T. Chebotarev, M. A. Adrianov, V. N. Konev, I. G. Labedev, V. I. Bagrova, A. V. Beznosikova, A. A. Kruglov, P. N. Petrov, and E. S. Smotvitskaya, *Sov. At. Energy* **23**, 1279 (1967).
144. K. H. J. Buschow and H. J. von Daal, *A.I.P. Conf. Proc.* **5**, 1464 (1972).
145. A. C. Gossard, V. Jaccarino, and J. H. Wernick, *Phys. Rev.* **128**, 1038 (1962).
146. A. R. Harvey, M. B. Brodsky, and W. J. Nellis, *Phys. Rev. B* **7**, 4137 (1973).
147. W. J. Nellis and M. B. Brodsky, *A.I.P. Conf. Proc.* **5**, Part 2, 1483 (1972).
148. N. Rivier and M. J. Zuckermann, *Phys. Rev. Lett.* **21**, 904 (1968).
149. A. B. Kaiser and S. Doniach, *Int. J. Magn.* **1**, 11 (1970).
150. M. Loewenhaupt, S. Horn, F. Steglich, E. Holland-Moritz and G. H. Lander, *J. Phys. (Paris)* **40**, C4-142 (1979).
151. M. B. Brodsky, *J. Phys. (Paris)* **40**, C4-147 (1979).
152. J. M. Fournier and J. Beille, *J. Phys. (Paris)* **40**, C4-145 (1979).
153. H. Armbrüster, W. Franz, W. Schlabitz, and F. Steglich, *J. Phys. (Paris)* **40**, C4-150 (1979).
154. J. Gal, Z. Hadari, U. Atzmony, E. R. Bauminger, I. Nowik, and S. Ofer, *Phys. Rev. B* **8**, 1901 (1973).
155. A. T. Aldred, B. D. Dunlap, D. J. Lam, and I. Nowik, *Phys. Rev. B* **10**, 1011 (1974).
156. M. B. Brodsky and R. J. Trainor, *J. Phys. (Paris)* **39**, C6-777 (1978).
157. J. G. Huber, M. B. Maple, and D. Wohlleben, *J. Magn. and Magn. Mat.* **1**, 58 (1975).
158. A. T. Aldred, *J. Magn. Magn. Mat.* **10**, 53 (1979).
159. G. H. Lander, A. T. Aldred, B. D. Dunlap, and G. K. Shenoy, *Physica* **86-88B & C**, 152 (1977).
160. A. T. Aldred, *J. Magn. Magn. Mat.* **10**, 42 (1979).
161. Y. Hamaguchi, N. Kunitomi, S. Komura, and M. Sakamoto, *J. Phys. Soc. Jpn* **17**, 398 (1962).
162. S. Lin and R. Ogilvie, *J. Appl. Phys.* **34**, 1372 (1963).
163. A. Misiuk, J. Mulak, A. Czopnik, and W. Trzebiatowski, *Bull. Acad. Pol. Sci., Ser. Sci. Chim.* **20**, 337 (1972).
164. A. T. Aldred, G. Cinader, D. J. Lam, and L. W. Weber, *Phys. Rev. B* **19**, 300 (1979).
165. K. Andres, D. Davidov, P. Dernier, F. Hsu, W. A. Reed, and G. J. Nieuwenhuys, *Solid State Commun.* **28**, 405 (1978).
166. N. Shamir, M. Melamud, H. Shaked, and M. Weger, *Physica* **94B**, 225 (1978).
167. R. J. Trainor, M. B. Brodsky, B. D. Dunlap, and G. K. Shenoy, *Phys. Rev. Lett.* **37**, 1511 (1976).

168. J. Grunzweig-Genossar, M. Kuznietz, and B. Meerovici, *Phys. Rev. B* **1**, 1958 (1970).
169. D. M. Gruen, *J. Chem. Phys.* **23**, 1708 (1955).
170. A. Misiuk, J. Mulak, and A. Czopnik, *Bull. Acad. Pol. Sci., Ser. Sci., Chim.* **20**, 891 (1972).
171. J. G. Huber and C. A. Luengo, *J. Phys. (Paris)* **39**, C6-781 (1978).
172. C. A. Luengo, J. G. Huber, R. W. McCallum, and M. B. Maple, in: *Low Temperature Physics − LT 14* (M. Krusius and M. Vuorio, eds.), Vol. 3, p. 192, North Holland, Amsterdam (1975).
173. H. J. van Daal, K. H. J. Buschow, P. B. van Aken, and M. H. van Maaren, *Phys. Rev. Lett.* **34**, 1452 (1975).
174. M. Coldea, I. Pop, W. E. Wallace, and K. S. V. L. Narasimhan, *Magn. Lett.* **1**, 11 (1976).
175. M. B. Brodsky and N. J. Bridges, *A.I.P. Conf. Proc.* **18**, 357 (1974).
176. A. V. Deryagin, A. V. Andreev, and V. A. Reimer, *Sov. Phys. JETP* **47**, 933 (1978).
177. M. B. Maple, J. G. Huber, B. R. Coles, and A. C. Lawson, *J. Low Temp. Phys.* **3**, 137 (1970).
178. V. O Struebing, H. H. Hill, and J. L. Smith, *Solid State Commun.* **25**, 1041 (1978).
179. E. F. Westrum, Jr., and F. Grønvold, *J. Phys. Chem. Solids* **23**, 39 (1962).
180. K. Gotoo, K. Naito, and S. Namura, *J. Phys. Chem. Solids* **26**, 1673 (1965).
181. L. M. Atlas, *J. Phys. Chem. Solids* **29**, 1349 (1968).
182. D. W. Osborne and E. F. Westrum, Jr., *J. Chem. Phys.* **21**, 1884 (1953).
183. W. M. Jones, J. Gordon, and E. A. Long, *J. Chem. Phys.* **20**, 695 (1952).
184. J. Faber, Jr., G. H. Lander, and B. R. Cooper, *Phys. Rev. Lett.* **35**, 1770 (1975).
185. M. J. M. Leask, L. E. Roberts, A. J. Walter, and W. P. Wolf, *J. Chem. Soc.* **1963**, 4788 (1963).
186. J. K. Dawson and M. W. Lister, *J. Chem. Soc. (London)* **1950**, 2181 (1950); **1952**, 5041 (1952).
187. O. G. Brandt and C. T. Walker, *Phys. Rev. Lett.* **18**, 11 (1967).
188. K. Aring and A. Sievers, *J. Appl. Phys.* **38**, 1496 (1967).
189. M. Daniel, *Phys. Lett.* **22**, 131 (1966).
190. J. Faber, Jr., *A.I.P. Conf. Proc.* **24**, 51 (1975).
191. P. J. Colwell, L. A. Rahn, and C. T. Walker, in: *Light Scattering in Solids* (Balkanski, Leite, and Porto, eds.), Int. Conf. Light Scattering, Campinas, Brasil (1975).
192. J. Schönes, *J. Appl. Phys.* **49**, 1463 (1978).
193. E. Slinowski and N. Elliott, *Acta Cryst.* **5**, 768 (1952).
194. J. B. Comly, *J. Appl. Phys.* **39**, 716 (1968).
195. W. Trzebiatowski and P. Selwood, *J. Am. Chem. Soc.* **72**, 4504 (1950).
196. J. K. Dawson and L. D. Roberts, *J. Chem. Soc.* **1956**, 78 (1956).
197. A. Arrott and J. Goldman, *Phys. Rev.* **108**, 948 (1957).
198. S. Siegel, *Acta Cryst.* **8**, 617 (1955).
199. B. Belbeoch, J. C. Boivineau, and P. Perio, *J. Phys. Chem. Solids* **28**, 1267 (1967).
200. J. Ross and D. J. Lam, *J. Appl. Phys.* **38**, 1451 (1967).
201. D. Cox and B. Frazer, *J. Phys. Chem. Solids* **28**, 1649 (1967).
202. L. Heaton, M. Mueller, and J. Williams, *J. Phys. Chem. Solids* **28**, 1651 (1967).
203. B. D. Dunlap, G. M. Kalvius, D. J. Lam, and M. B. Brodsky, *J. Phys. Chem. Solids* **29**, 1365 (1968).
204. *Proceedings 2nd International Conference on the Electronic Structure of the Actinides* (J. Mulak, W. Suski, and R. Troć, eds.), Wrocław (1977).

REFERENCES

205. M. Bacmann, E. F. Bertaut, and A. Blaise, *Compt. Rend. Acad. Sc. Paris, B* **266**, 45 (1968).
206. M. Bacmann, R. Chevalier, E. F. Bertaut, G. Roult, and M. Belakhovsky, *Compt. Rend. Acad. Sc. Paris B* **267**, 518 (1968).
207. M. Bacmann, E. F. Bertaut, A. Blaise, R. Chevalier, and G. Roult, *J. Appl. Phys.* **40**, 1131 (1969).
208. L. D. Roberts and R. B. Murray, *Phys. Rev.* **100**, 650 (1955).
209. J. K. Dawson, (a) *J. Chem. Soc.* **1951**, 429 (1951); (b) *J. Chem. Soc.* **1951**, 2889 (1951); (c) *J. Chem. Soc.* **1952**, 1185 (1952).
210. M. J. M. Leask, D. W. Osborne, and W. P. Wolf, *J. Chem. Phys.* **34**, 2090 (1961).
211. J. H. Burns, D. W. Osborne, and E. F. Westrum, Jr., *J. Chem. Phys.* **33**, 387 (1960).
212. S. P. Gabuda, L. G. Falaleeva, and Yu. V. Gagarisnkii, *Phys. Status Solidi* **33**, 435 (1969).
213. J. Stone and E. R. Jones, *J. Chem. Phys.* **54**, 1713 (1971).
214. J. Mulak and Z. Zolnierek, private communication.
215. M. E. Hendricks, E. R. Jones, Jr., J. A. Stone, and D. G. Karraker, *J. Chem. Phys.* **55**, 2993 (1971).
216. W. Trzebiatowski, in Ref. 151, p. 23.
217. A. F. Leung, *J. Phys. Chem. Solids* **38**, 529 (1977).
218. E. Soulie, *J. Phys. Chem. Solids* **39**, 695 (1978).
219. F. L. Carter, P. Wolfers, and G. Fillion, *J. Chem. Soc. Dalton Trans.* 1686 (1978).
220. B. Kanellakopulos, C. Aderhold, E. Dornberger, W. Muller, and R. D. Baybarz, *Radiochim. Acta* **25**, 89 (1978).
221. A. Blaise, R. Lagnier, J. Mulak, and Z. Zolnierek, *J. Phys. (Paris)* **40**, C4-176 (1979).
222. W. A. Hargreaves, *Phys. Rev.* **156**, 331 (1967).
223. R. A. Satten, C. L. Schreiber, and E. Y. Wong, *J. Chem. Phys.* **42**, 162 (1965).
224. G. A. Candela, C. A. Hutchinson, Jr., and W. B. Lewis, *J. Chem. Phys.* **30**, 246 (1959).
225. D. Johnston, R. Satten, and E. Wong, in: *Optical Properties of Ions in Crystals* (H. Crosswhite and H. Moos, eds.), p. 429, Interscience, London (1967); and *J. Chem. Phys.* **44**, 687 (1966).
226. N. Edelstein, H. F. Mollet, W. C. Easley, and R. J. Mehlhorn, *J. Chem. Phys.* **51**, 3281 (1969).
227. M. M. Abraham, L. A. Boatner, C. B. Finch, and R. W. Reynolds, *Phys. Rev. B* **3**, 2864 (1971).
228. J. G. Conway and R. Rajnak, *J. Chem. Phys.* **44**, 348 (1966).
229. J. G. Conway, *J. Chem. Phys.* **41**, 904 (1964).
230. W. T. Carnall and B. G. Wybourne, *J. Chem. Phys.* **40**, 3428 (1964).
231. K. K. Sharma and J. O. Artman, *J. Chem. Phys.* **50**, 1241 (1969).
232. L. P. Varga, M. J. Reisfeld, and L. B. Asprey, *J. Chem. Phys.* **53**, 1 (1970).
233. J. B. Gruber, W. R. Cochran, J. G. Conway, and A. T. Nicol, *J. Chem. Phys.* **45**, 1423 (1966).
234. L. P. Varga, J. D. Brown, M. J. Reisfeld, and R. D. Cowan, *J. Chem. Phys.* **52**, 4233 (1970).
235. D. J. Lam and F. Y. Fradin, *Phys. Rev. B* **1**, 238 (1974).
236. P. F. McDonald, *Phys. Rev.* **177**, 447 (1969).
237. C. M. Bowden, H. C. Meyer, P. F. McDonald, and J. D. Stettler, *J. Phys. Chem. Solids* **30**, 1535 (1969).
238. J. C. Eisenstein and M. Pryce, *Proc. R. Soc. (London) A* **255**, 181 (1960).

239. M. Genet, P. Delamoye, N. Edelstein, and J. Conway, *J. Chem. Phys.* **67**, 1620 (1977).
240. P. Delamoye, S. Hubert, M. Hussonais, J. C. Krupa, M. Genet, R. Guillamont, N. Edelstein, and J. Conway, *J. Phys. (Paris)* **40**, C4-173 (1979).
241. J. E. Bray, *Phys. Rev. B.* **18**, 2973 (1978).
242. R. P. Richardson and J. B. Gruber, *J. Chem. Phys.* **65**, 256 (1972).
243. R. P. Richardson and J. B. Gruber, *J. Chem. Phys.* **62**, 1296 (1975).
244. J. G. Conway, *J. Chem. Phys.* **40**, 2504 (1964).
245. P. G. Pappalardo, W. T. Carnall, P. R. Fields, *J. Chem. Phys.* **51**, 1182 (1969).
246. M. Kuznietz, G. H. Lander, and F. P. Campos, *J. Phys. Chem. Solids* **30**, 1642 (1969).
247. C. E. Olsen and W. C. Koehler, *J. Appl. Phys.* **40**, 1135 (1969).
248. D. J. Lam, B. D. Dunlap, A. R. Harvey, M. H. Mueller, A. T. Aldred, I. Nowik, and G. H. Lander, discourse, *Int. Conf. on Magnetism*, Moscow (1973) (unpublished).
249. A. T. Aldred, B. D. Dunlap, D. J. Lam, G. H. Lander, M. H. Mueller, and I. Nowik, *Phys. Rev. B* **11**, 530 (1975).
250. N. Nereson, C. Olsen, and G. Arnold, *J. Appl. Phys.* **37**, 4575 (1966).
251. J. Mulak and Z. Zolnierek, p. 125 in Ref. 204.
252. V. I. Checkernikov, T. M. Shavishvili, V. A. Pletyushkin, and V. K. Slavyanskikh, *Sov. Phys. JETP* **28**, 81 (1969).
253. M. Steinitz and J. Grunzweig-Genossar, *J. Phys. (Paris)* **40**, C4-34 (1979).
254. P. Burlet, S. Quezel, J. Rossat-Mignod, H. Bartholin, and O. Vogt, 11th Journées des Actinides, Jesolo Lido (1981), p. 154.
255. *Mössbauer Isomer Shifts*, (G. K. Shenoy and F. E. Wagner, eds.), North-Holland (1978).
256. W. Trzebiatowski, in: *Ferromagnetic Materials* (E. P. Wohlfarth, ed.) North-Holland (1980), Vol. 1, ch. 5.
257. O. Vogt, P. Wachter, and H. Bartholin, *Physica* **102 B+C**, 226 (1980).
258. O. Vogt, *Physica* **102 B+C**, 206 (1980).
259. J. Rossat-Mignod, P. Burlet, S. Quezel, and O. Vogt, *Physica* **102 B+C**, 237 (1980).
260. R. Troć, J. Leciejewicz, and G. H. Lander, *J. Magn. Magn. Mat.* **21**, 173 (1980).
261. R. Troć, 11th Journées des Actinides, Jesolo Lido (1981), p. 120.
262. J. Rossat-Mignod, P. Burlet, H. Bartholin, R. Tchapoutian, O. Vogt, C. Vettier, and R. Lagnier, *Physica* **102 B+C**, 177 (1980).
263. J. M. Fournier, J. Beille, A. Boeuf, C. Vettier, and A. Wedgwood, *Physica* **102 B+C**, 282 (1980).
264. P. R. Norton, R. L. Tapping, D. K. Creber, and W. J. L. Buyers, *Phys. Rev.* **B21**, 2572 (1980).
265. W. J. L. Buyers, A. F. Murray, T. M. Holden, E. C. Svensson, P. de V. Du Plessis, G. H. Lander, and O. Vogt, *Physica* **102 B+C**, 291 (1980).
266. J. Schönes and O. Vogt, *J. Phys. (Paris)* **40**, C4-38 (1979).
267. P. Burlet, S. Quezel, J. Rossat-Mignod, O. Vogt, and G. H. Lander, *Physica* **102 B+C**, 271 (1980).
268. F. Hulliger and O. Vogt, *Physica* **102 B+C**, 316 (1980).
269. Y. Baer, *Physica* **102 B+C**, 104 (1980).
270. M. Erbudak and F. Meier, *Physica* **102 B+C**, 134 (1980).
271. M. Erbudak and J. Keller, *Z. Phys.* **B23**, 281 (1979).
272. J. Schönes, *Physica* **102 B+C**, 45 (1980).
273. A. Delapalme, G. Busch, O. Vogt, and G. H. Lander, *J. de Phys. (Paris)* C 4- 74 (1979).

REFERENCES

274. A. Blaise, D. Damien, and W. Suski, *Solid State Commun.* **37**, 659 (1981).
275. G. H. Lander, and D. J. Lam, *Phys. Rev. B* **14**, 4064 (1976).
276. M. H. Mueller, G. H. Lander, H. A. Hoff, H. W. Knott, and J. F. Reddy, *J. de Phys. (Paris) C* **4**- 68 (1979).
277. R. Lasser, J. C. Fuggle, M. Beyss, M. Campagna, F. Steglich, and F. Hulliger, *Physica* **102 B+C**, 360 (1980).
278. R. J. Trainor, M. B. Brodsky, and G. S. Knapp, *Plutonium 1975 and Other Actinides*, North-Holland (1975), p. 475.
279. J. R. Naegele, L. Manes, J. C. Spirlet, L. Pellegrini, and J. M. Fournier, *Physica* **102 B+C**, 122 (1980).
280. J. Moser, J. Gal, W. Potzel, G. Wortmann, G. M. Kalvius, B. D. Dunlap, D. J. Lam, and J. C. Spirlet, *Physica* **102 B+C**, 199 (1980).
281. W. D. Schneider and C. Laubschat, *Phys. Lett.* **75A**, 407 (1980).
282. W. D. Schneider and C. Laubschat, *Physica* **102 B+C**, 111 (1980).
283. R. Siemann and B. Cooper, *Phys. Rev. B* **20**, 2869 (1979).
284. B. W. Veal, *J. Phys. (Paris) C* **4**- 163 (1979).
285. A. Delapalme, M. Forte, J. M. Fournier, J. Rebizant, and J. C. Spirlet, *Physica* **102 B+C**, 171 (1980).
286. P. Erdös, G. Solt, Z. Żołnierek, A. Blaise, and J. M. Fournier, *Physica* **102 B+C**, 164 (1980).
287. A. Murasik, P. Fischer, and W. Szczepaniak, *J. Phys. C* **14**, 1847 (1981).
288. A. Murasik, A. Furrer, and W. Szczepaniak, *Solid State Commun.* **33**, 1217 (1980).
289. E. R. Jones, Jr., M. E. Hendricks, J. A. Stone, and D. G. Karraker, *J. Chem. Phys.* **60**, 2088 (1974).
290. G. Solt and P. Erdös, *Phys. Rev. B* **22**, 4718 (1980).
291. P. Delamoye, C. K. Malek, S. Hubert, J. C. Krupa, M. Genet, and R. Guillamont, 11th Journées des Actinides, Jesolo Lido (1981), p. 183.
292. J. Jensen and P. Bak, H. C. Orsted Institute, Preprint (Dec. 1980).
293. M. J. Mortimer, AERE Report R 8852 (1977).
294. H. R. Haines, R. O. A. Hall and J. A. Lee, p. 349 in Ref. 204.
295. D. G. Martin *et al.*, AERE Report HL 76, 2599 (1976).
296. A. Boeuf, J. M. Fournier, L. Manes, J. Rebizant, and F. Rustichelli, 11th Journées des Actinides, Jesolo Lido (1981), p. 233.
297. R. O. A. Hall, M. J. Mortimer, and G. H. Lander, 11th Journées des Actinides, Jesolo Lido (1981), p. 156.
298. H. H. Hill, *Nucl. Mat.* **17**, part 1, 2 (1970).
299. H. Grohs, H. Höchst, P. Steiner, S. Hufner, and K. H. J. Buschow, *Sol. State Commun.* **33**, 573 (1980).
300. J. Faber, Jr., and G. H. Lander, *Phys. Rev. B* **14**, 1151 (1976).
301. W. T. Carnall, H. Crosswhite, H. M. Crosswhite, J. P. Hessler, N. Edelstein, J. G. Conway, G. V. Shalimoff, and R. Sarup, *J. Chem. Phys.* **72**, 5089 (1980).
302. S. P. Cook, Ph.D. Thesis, The John Hopkins University, Baltimore, MD (1959).
303. H. M. Crosswhite, H. Crosswhite, W. T. Carnall, and A. P. Paszek, *J. Chem. Phys.* **72**, 5103 (1980).
304. C. J. Ballhausen, *Introduction to Ligand Field Theory*, McGraw-Hill, New York (1962).
305. R. D. Cowan and C. D. Griffin, *J. Opt. Soc. Am.* **66**, 1010 (1976).
306. R. Allen and M. S. S. Brooks, *J. Phys. (Paris) C* **4**- 19 (1979).
307. A. J. Freeman and R. E. Watson, *Phys. Rev.* **127**, 2058 (1962).
308. C. J. Lenander, *Phys. Rev.* **130**, 1033 (1963).

309. P. Erdös and P. Rudra, *Bull. Am. Phys. Soc.* **II13**, 1395 (1968); and Florida State University report (unpublished).
310. A. J. Freeman and R. E. Watson, *Phys. Rev. A* **135**, 1209 (1964); **139**, 1660 (1965); **156**, 251 (1967).
311. R. E. Watson and A. J. Freeman, *Phys. Rev.* **133**, A1571 (1964); **139**, A1606 (1965); **156**, 251 (1967).
312. K. R. Lea, M. J. M. Leask, and W. P. Wolf, *J. Phys. Chem. Solids* **23**, 1381 (1962).
313. S. K. Chan and D. J. Lam, in Ref. 23, Vol. 1, p. 1.
314. K. W. H. Stevens, *Proc. Phys. Soc. (London) A* **65**, 209 (1952).
315. E. Fick and G. Joos, *Kristallspektren*, Handbuch der Physik, Vol. 28, Spektroskopie II (S. Flügge, ed.) Springer-Verlag, Berlin (1957).
316. P. Erdös and H. A. Razafimandimby, *Helv. Phys. Acta* **51**, 12 (1978).
317. B. R. Judd, *Operator Techniques in Atomic Spectroscopy*, McGraw-Hill, New York (1963).
318. C. W. Nielson and G. F. Koster, *Spectroscopic Coefficients for the p, d and f Configurations*, MIT Press, Cambridge, Mass. (1963).
319. K. Rajnak and B. G. Wybourne, *Phys. Rev.* **132**, 280 (1963).
320. B. R. Judd, H. M. Crosswhite, and H. Crosswhite, *Phys. Rev.* **169**, 130 (1968).
321. Landolt-Boernstein, *Numerical Data and Functional Relationships in Science and Technology*, Neue Serie I,3, Numerical Tables for 3J-, 6J-, 9J-Symbols, F-, and Γ-coefficients, (K. H. Hellwege, H. Schopper, and H. Appel, eds.), Springer-Verlag, Berlin (1968).
322. R. M. Sternheimer, *Phys. Rev.* **146**, 140 (1966).
323. P. Erdös and J. Kang, unpublished.
324. F. C. Von der Lage and H. A. Bethe, *Phys. Rev.* **71**, 612 (1947).
325. H. G. Hecht and J. B. Gruber, *J. Chem. Phys.* **60**, 4872 (1971).
326. J. B. Gruber and H. G. Hecht, *J. Chem. Phys.* **60**, 1352 (1974).
327. P. Erdös and J. Kang, *Phys. Rev. B* **6**, 3393 (1972).
328. C. J. Lenander, private communication.
329. A. R. Edmonds, *Angular Momentum in Quantum Mechanics*, Princeton U. Press, Princeton (1960).
330. B. R. Judd, *J. Chem. Phys.* **66**, 3163 (1977).
331. M. E. Rose, *Elementary Theory of Angular Momentum*, John Wiley & Sons, New York (1967).
332. R. H. Lemmer and J. E. Lowther, *J. Phys. C* **11**, 1145 (1978).
333. G. H. Lander and W. G. Stirling, *Phys. Rev. B* **21**, 436 (1980).
334. R. Troć and D. J. Lam, *Phys. Status Solidi B* **65**, 317 (1974).
335. S. K. Sinha, G. H. Lander, S. M. Shapiro, and O. Vogt, *Phys. Rev. B* **23**, 4556 (1981).
336. C. E. T. Gonçalves da Silva, *J. Magn. Magn. Mat.* **15-18**, 945 (1980).
337. P. W. Anderson, in: *Magnetism* (G. Rado and H. Suhl, eds.) Vol. I, Academic Press, New York (1963).
338. R. J. Birgeneau, M. T. Hutchings, J. M. Baker, and J. D. Riley, *J. Appl. Phys.* **40**, 1070 (1969).
339. M. E. Lines and E. D. Jones, *Phys. Rev.* **139**, A1313 (1965).
340. D. S. Rodbell, *Phys. Rev. Lett.* **11**, 10 (1963).
341. M. A. Ruderman and C. Kittel, *Phys. Rev.* **96**, 99 (1954).
342. J. M. Robinson and P. Erdös, *Phys. Rev. B* **6**, 3337 (1972).
343. R. J. Elliot and F. A. Wedgwood, *Proc. Phys. Soc. (London)* **81**, 846 (1963); **84**, 63 (1964).

REFERENCES

344. H. Miwa, *Proc. Phys. Soc. (London)* **85**, 1197 (1965).
345. B. Coqblin and J. R. Schrieffer, *Phys. Rev.* **185**, 847 (1969).
346. B. R. Cooper and R. Siemann, in: *Crystalline Electric Field and Structural Effects in f-Electron Systems* (J. E. Crow, R. P. Guertin, and T. W. Mihalisin, eds.), p. 241, Plenum, New York (1980).
347. B. Dunlap, *J. Appl. Phys.* **40**, 1495 (1969).
348. G. Solt and P. Erdös, *J. Magn. Magn. Mat.* **15-18**, 57 (1980).
349. G. Solt and P. Erdös, in: *Crystalline Electric Field and Structural Effects in f-Electron Systems* (J. E. Crow, R. P. Guertin, and T. W. Mihalisin, eds.), p. 275, Plenum, New York (1980).
350. Z. Żołnierek, G. Solt, and P. Erdös, *J. Phys. Chem. Sol.* **42**, 773 (1981).
351. C. Long, Thesis, Florida State University (1970).
352. M. Blume, *Phys. Rev.* **141**, 517 (1966).
353. R. Jullien, E. Galleani d'Agliano, and B. Coqblin, *Phys. Rev. B* **6**, 2139 (1972).
354. R. Jullien and B. Coqblin, *Phys. Rev. B* **8**, 5263 (1973).
355. D. D. Koelling and A. J. Freeman, *Phys. Rev.* **7**, 4454 (1973).
356. D. D. Koelling and A. J. Freeman, *A.I.P. Conf.* **10**, 1300 (1973).
357. H. L. Davis, in: *Plutonium 1970 and the Other Actinides* (W. N. Miner, ed.); *Nuclear Metallurgy* **17**, Part I, p. 209 (1970).
358. R. Jullien, M. T. Béal-Monod, and B. Coqblin, *Phys. Rev. B* **9**, 1441 (1974).
359. M. S. S. Brooks and D. Glotzel, *Physica* **102 B**, 51 (1980).
360. M. S. S. Brooks, *J. de Phys. (Paris) C* 4- 155 (1979).
361. M. B. Brodksy, *Phys. Rev. B* **9**, 1381 (1974).
362. N. Rivier and V. Zlatić, *J. Phys. F* **2**, L99 (1972).
363. P. W. Anderson, *Phys. Rev.* **124**, 41 (1961).
364. J. Friedel, *Nuovo Cimento Suppl.* **7**, 287 (1958).
365. S. Doniach, in Ref. 23, Vol. 2, Chap. 2, p. 51.
366. P. Erdös and J. M. Robinson, *A.I.P. Conf. Proc.* **10**, 1070 (1973).
367. L. M. Falicov and J. C. Kimball, *Phys. Rev. Lett.* **22**, 297 (1969).
368. R. Ramirez, L. M. Falicov, and J. C. Kimball, *Phys. Rev. B* **2**, 3383 (1970).
369. R. Ramirez and L. M. Falicov, *Phys. Rev. B* **3**, 2425 (1971).
370. L. M. Falicov, C. E. T. Gonçalves da Silva, and B. A. Huberman, *Solid State Commun.* **10**, 455 (9172).
371. C. E. T. Gonçalves da Silva and L. M. Falicov, *J. Phys. C* **5**, 63 (1972).
372. J. Grunzweig-Genossar and J. W. Cahn, *Int. J. Magn.* **4**, 193 (1970).
373. J. Grunzweig-Genossar, private communication.
374. S. B. Haley and P. Erdös, *Phys. Rev. B* **5**, 1106 (1972).
375. H. Hubbard, *Proc. Phys. Soc. A* **277**, 237 (1964); *A* **285**, 542 (1965).
376. R. Stevenson, *Multiplet Structure of Atoms and Molecules*, W. B. Saunders, Philadelphia (1965).
377. Y. P. Irkhin, *Sov. Phys. JETP* **39**, 490 (1974).
378. J. C. Slater and G. F. Koster, *Phys. Rev.* **91**, 1498 (1954).
379. J. H. van Vleck, *Rev. Mod. Phys.* **25**, 220 (1953).
380. L. L. Hirst, *A.I.P. Conf. Proc.* **24**, 11 (1975).
381. S. V. Vonsovskii, *Magnetism*, Vol. 2, p. 661, John Wiley and Sons, Jerusalem (1974).
382. J. M. Robinson, *A.I.P. Conf. Proc.* **29**, 319 (1976).
383. K. Rajnak, *Phys. Rev. A* **14**, 1979 (1976).
384. C. Y. Huang, R. J. Lastowski, C. E. Olsen, and J. L. Smith, *J. Phys. (Paris) C* 4- 26 (1979).

385. J. M. Robinson, *Phys. Rep.* **51**, 1–62 (1979).
386. U. Fano and G. Racah, *Irreducible Tensorial Sets*, Academic Press, New York (1959).
387. W. Voigt, *Lehrbuch der Kristallphysik*, Teubner, Leipzig (1910); see also M. Born and K. Huang, *Dynamical Theory of Crystal Lattices*, p. 129, Oxford, London (1954).
388. J. M. Fournier, J. Beille, and C. H. de Novion, *J. Phys. (Paris)* **40**, C4-32 (1979).
389. G. T. Meaden, *Electrical Resistance of Metals*, Heywood, London (1966).
390. C. E. Olsen and R. O. Elliott, *Phys. Rev.* **139**, A 437 (1965).
391. J. M. Robinson, *A.I.P. Conf. Proc.* **34**, 189 (1976).
392. J. M. Robinson, in: *Valence Instabilities and Related Narrow Band Phenomena* (R. D. Parks, ed.), p. 423, Plenum Press, New York (1977).
393. E. Franceschi and G. L. Olcese, *Phys. Rev. Lett.* **22**, 1299 (1969).
394. A. Chatterjee, A. K. Singh, A. Jayaraman, *Phys. Rev. B* **6**, 2285 (1972).
395. H. Bjerrum Møller, S. M. Shapiro, and R. J. Birgeneau, *Phys. Rev. Lett.* **39**, 1021 (1977).
396. J. L. Prather, National Bureau of Standards Monograph 19, U.S. Govt. Printing Office, Washington D.C. (1961).
397. T. Penney and F. Holtzberg, *Phys. Rev. Lett.* **34**, 322 (1975).
398. T. G. Godfrey, W. Fulkerson, T. Kollie, J. Moore, and D. Mc. Elroy, *J. Am. Ceram. Soc.* **48**, 297 (1965).
399. B. R. Cooper and R. Siemann, *Phys. Rev. B* **19**, 2645 (1979).
400. F. Greuter, E. Hauser, P. Oelhafen, H. J. Güntherodt, B. Reihl, and O. Vogt, *Physica* **102 B+C**, 117 (1980).
401. J. Mulak and Z. Żołnierek, *Solid State Commun.* **26**, 275 (1978).
402. P. Wolfers, G. Fillion, M. Bacmann, and H. Noel, *J. Phys. (Paris)* **37**, 233 (1976).
403. B. W. Veal, D. J. Lam, H. Diamond, and H. R. Hoekstra, *Phys. Rev. B* **15**, 2929 (1977).
404. B. W. Veal, D. J. Lam, H. R. Hoekstra, H. Diamond, and W. T. Carnall, p. 145 in Ref. 204.
405. W. B. Lewis, J. B. Mann, D. A. Liberman, and D. T. Cromer, *J. Chem. Phys.* **53**, 809 (1970).
406. J. J. Huntzicker and E. F. Westrum, Jr., *J. Chem. Thermodyn.* **3**, 61 (1971).
407. *Third International Conference on the Electronic Structure of the Actinides*, Grenoble, France, Supplément au Journal de Physique (Paris), Colloque No. 4 (1979).
408. *Proceedings of the International Conference on the Physics of Actinides and Related 4f Materials*, Zurich, Switzerland, *Physica* **102 B+C** (1980).
409. *Proc. Int. Conf. on The Physics of Actinides and Related 4f Materials* (P. Wachter, ed.), North Holland, Amsterdam (1980).
410. J. C. Spirlet, E. Bednarczyk, C. Rijkaboer, and W. Muller, 12èmes Journées des Actinides, Orsay, France (1982), (unpublished).
411. P. Burlet, S. Quezel, J. Rossat-Mignod, and R. Horyn, 12èmes Journées des Actinides, Orsay, France (1982), (unpublished).
412. M. Loewenhaupt, G. H. Lander, A. Murani, A. Murasik, discourse, IV Int. Conf. on Crystal Field and Structural Effects in *f*-Electron Systems, Wrocław (1981).
413. U. Benedict, J. C. Spirlet, J. S. Olsen, S. Steenstrup, and L. Gerward, 12èmes Journées des Actinides, Orsay, France (1982), (unpublished).
414. H. Noël, 12èmes Journées des Actinides, Orsay, France (1982), (unpublished).
415. J. Schoenes and O. Vogt, 12èmes Journées des Actinides, Orsay France (1982), unpublished.

REFERENCES

416. A. Blaise, J. M. Fournier, E. Roudault, and A. Wojakowski, 12èmes Journées des Actinides, Orsay, France (1982), (unpublished).
417. R. Baptist, J. Chayrouse, D. Courteix, L. Heintz, D. Damien, and A. Wojakowski, 12èmes Journées des Actinides, Orsay, France (1982), (unpublished).
418. P. Burlet, S. Quezel, J. Rossat-Mignod, O. Vogt, and H. Bartholin, 12èmes Journées des Actinides, Orsay, France (1982), (unpublished).
419. G. Amoretti and A. Blaise, 12èmes Journées des Actinides, Orsay, France (1982), (unpublished).
420. M. Bogé, J. Chapper, and A. Wojakowski, 12èmes Journées des Actinides, Orsay, France (1982), (unpublished).
421. K. G. Gurtovoi, A. S. Lagutin, R. Z. Levitin, and V. I. Ozhogin, discourse, IV Int. Conf. on Crystal-Field and Structure Effects in f-Electron Systems, Wroclaw (1981).
422. P. J. Markowski, S. Kunii, T. Suzuki, Z. Henkie, T. Kasuya, discourse, IV Int. Conf. on Crystal-field and Structural Effects in f-Electron Systems, Wroclaw (1981).
423. K. Takegahara, A. Yanase, and T. Kasuya, discourse, IV Int. Conf. on Crystal Field and Structural Effects in f-Electron Systems, Wroclaw (1981).
424. J. Gal, S. Fredo, M. Kuznietz, W. Potzel, L. Asch, and G. M. Kalvius, 12èmes Journées des Actinides, Orsay, France (1982), (unpublished).
425. J. Brunner, M. Erbudak, and F. Hulliger, 12èmes Journées des Actinides, Orsay, France (1982), (unpublished).
426. Z. Żołnierek, discourse, IV Int. Conf. on Crystal-Field and Structural Effects in f-Electron Systems, Wroclaw (1981).
427. P. Fischer and A. Murasik, 12èmes Journées des Actinides, Orsay, France (1982), (unpublished).
428. R. G. Haïre, S. E. Nave, and P. G. Huray, 12èmes Journées des Actinides, Orsay, France (1982), (unpublished).
429. J. K. Lang and Y. Baer, *Rev. Sci. Instr.* **50**, 221 (1979).
430. Y. Baer and J. Schoenes, *Solid State Commun.* **33**, 885 (1980).
431. G. Chovat and R. Baptist, *J. Electron Spectrosc. and Relat. Phenom.* **24**, 255 (1981).
432. R. Baptist, Thèse, Université Scientifique et Médicale de Grenoble, France (March, 1982).
433. J. P. Hessler and W. T. Carnall, *Lanthanide and Actinide Chemistry and Spectroscopy*, N. M. Edelstein, ed., ACS Symp. Series **131**, 349 (1980).
434. J. Rossat-Mignod, S. Quezel, P. Burlet, A. Blaise, J. M. Fournier, D. A. Damien, and A. Wojakowski, Proc. 11èmes Journées des Actinides, Jesolo Lido (1981), p. 143.
435. A. Delapalme, A. Blaise, J. M. Fournier, D. Damien, and J. P. Charvillat, Proc. 11èmes Journées des Actinides, Jesolo Lido (1981), p. 144.
436. J. Brunner, M. Erbudak, and F. Hulliger, Proc. 11èmes Journées des Actinides, Jesolo Lido (1981), p. 175.
437. Z. Henkie, R. Maslanka, and M. Konczykowski, Proc. 11èmes Journées des Actinides, Jesolo Lido (1981), p. 204.
438. J. Schoenes, M. Küng, and Z. Henkie, Proc. 11èmes Journées des Actinides, Jesolo Lido (1981), p. 227.
439. R. Siemann and B. Cooper, *Phys. Rev. Lett.* **44**, 1015 (1980).
440. D. Yang and B. R. Cooper, Proc. IV Int. Conf. on Crystal-Field and Structural Effects in f-Electron Systems, Wroclaw, Poland (1981).

REFERENCES

441. P. Thayamballi and B. R. Cooper, private communication.
442. D. Yang and B. R. Cooper, *J. Appl. Phys.* **52**, 2234 (1981).
443. H. A. Razafimandimby and P. Erdös, *Z. Phys.* B **46**, 193 (1982).
444. W. E. Pickett, A. J. Freeman, and D. D. Koelling, *Phys. Rev. B* **23**, 1266 (1981).
445. B. Frick, J. Schoenes, O. Vogt, and J. W. Allen, *Solid State Comm.* **42**, 331 (1982).
446. J. Schoenes and K. Andres, *Solid State Comm.* **42**, 359 (1982).
447. J. Rossat-Mignod, P. Burlet, S. Quezel, O. Vogt, and H. Bartholin, Proc. IV Int. Conf. on Crystal-Field and Structural Effects in f-Electron Systems, Wroclaw, Poland (1981).
448. P. de V. du Plessis, C. F. van Doorn, N. J. S. Grobler, and D. C. van Delden, *J. Phys.* C **15**, 1525 (1982).
449. A. T. Aldred, P. de V. du Plessis, and G. H. Lander, *J. Magn. Magn. Mat.* **20**, 236 (1980).
450. C. F. van Doorn, private communication (1982).
451. G. L. Squires, *Introduction to the Theory of Thermal Neutron Scattering*, Cambridge University Press, Cambridge (1978).
452. P. Monachesi and F. Weling, private communication.

Author Index

The numbers refer to the list of References.

Abraham, M. M., 227
Adachi, H., 115
Aderhold, C., 220
Adrianov, M. A., 143
Aldred, A. T., 90, 91, 155, 158, 159, 160, 164, 249, 449
Allen, J. W., 445
Allen, R., 306
Allen, S. J., Jr., 14
Amoretti, G., 419
Anderson, K. D., 4, 53, 54
Anderson, P. W., 337, 363
Andreev, A. V., 176
Andres, K., 165, 446
Appel, H., 321
Aring, K., 188
Arko, A. J., 18
Armbrüster, H., 153
Arnold, G., 250
Arrot, A., 197
Artman, J. O., 231
Asch, L., 424
Asprey, L. B., 232
Atlas, L. M., 181
Atzmony, U., 154

Bacmann, M., 205, 206, 207, 402
Baer, Y., 269, 429, 430
Bagrova, V. I., 143
Bak, P., 292
Baker, J. M., 338
Ballestracci, R., 111
Ballhausen, C. J., 304
Baptist, R., 417, 431, 432

Bartholin, H., 62, 254, 257, 262, 418, 447
Barykin, M. E., 139
Baskin, Y., 51
Bates, L. F., 47
Bauminger, E. R., 154
Baybarz, R. D., 220
Bazan, C., 120, 134
Beal-Monod, M. T., 358
Bean, C. P., 15
Bednarczyk, E., 26, 410
Beille, J., 152, 263, 388
Belakhovsky, M., 206
Belbeoch, B., 199
Benedict, U., 93, 413
Bertaut, E. F., 111, 205, 206, 207
Bethe, H. A., 324
Betterton, J. O., 65
Beyss, M., 277
Beznosikova, A. V., 143
Bieganski, Z., 132
Bielow, K. P., 138
Birgeneau, R. J., 338, 395
Bjerrum-Møller, H., 395
Blaise, A., 95, 205, 207, 221, 274, 286, 416, 419, 434, 435
Blume, M., 352
Boatner, L. A., 227
Boeuf, A., 263, 296
Bogé, M., 420
Boivineau, J. C., 199
Born, M., 387
Bowden, C. M., 237
Brandt, O. G., 187
Bray, J. E., 241
Bridges, N. J., 175

AUTHOR INDEX

Brodsky, M. B., 7, 18, 57, 146, 147, 151, 156, 167, 175, 203, 278, 361
Brooks, M. S. S., 306, 359, 360
Brunner, J., 425, 436
Buhrer, C. F., 29
Burlet, P., 103, 254, 259, 262, 267, 411, 418, 434, 447
Burns, J. H., 211
Busch, G., 28, 62, 273
Buschow, K. H. J., 144, 173, 299
Buyers, W. J. L., 71, 264, 265

Cahn, J. W., 372
Calestani, G., 34
Campagna, M., 277
Campos, F. P., 246
Candela, G. A., 224
Carnall, W. T., 230, 245, 301, 303, 404, 433
Carr, S. L., 1
Carter, F. L., 219
Cater, E. D., 49
Chan, S. K., 313
Chapper, J., 420
Charvillat, J. P., 435
Chatterjee, A., 394
Chayrouse, J., 417
Chebotarev, N. T., 143
Chechernikov, V. I., 74, 139, 252
Chevalier, R., 206, 207
Chovat, G., 431
Cinader, G., 164
Ciszewski, R., 110, 128
Cochran, W. R., 233
Coldea, M., 174
Coles, B. R., 45, 177
Colwell, P. J., 191
Comly, J. B., 194
Conway, J. G., 228, 229, 233, 239, 240, 244, 301
Cook, S. P., 302
Cooper, B. R., 59, 102, 124, 184, 283, 346, 399, 439, 440, 441, 442
Coqblin, B., 345, 353, 354, 358
Costa, P., 44, 92
Counsell, J. F., 8, 46, 100
Courteix, D., 417
Cowan, R. D., 234, 305
Cowley, R. A., 25
Cox, D. E., 10, 11, 201
Crangle, J., 52

Creber, D. K., 264
Cromer, D. T., 405
Crosswhite, H., 301, 303, 320
Crosswhite, H. M., 301, 303, 320
Curry, N. A., 55, 63
Czopnik, A., 121, 125, 170

Damien, D. A., 31, 127, 274, 417, 434, 435
Daniel, M., 189
Darby, J. B., Jr., 23
Davidov, D., 165
Davidson, J. R., 57
Davis, H. L., 65, 357
Dawson, J. K., 186, 196, 209
Delamoye, P., 239, 240, 291
Delapalme, A., 273, 285, 435
Dell, R. M., 8, 46
de Novion, C. H., 43, 50, 127, 141, 388
Dernier, P., 165
Deryagin, A. V., 176
Desclaux, J. P., 80
Diamond, H., 403, 404
Didchenko, R., 66
Dmitrewskij, A. S., 138
Dobrinski, L., 131
Dolling, G., 25
Doniach, S., 149, 365
Dornberger, E., 220
Downie, D. B., 116
Duczmal, M., 119
Dufour, C., 93
Dunlap, B. D., 40, 57, 90, 91, 142, 155, 159, 167, 203, 248, 249, 280, 347
du Plessis, P. de V., 24, 67, 265, 448, 449

Easley, W. C., 226
Eastman, D., 84
Edelstein, N., 226, 239, 240, 301
Edmonds, A. R., 329
Eib, W., 85, 86
Eisenstein, J. C., 238
Ellert, G. V., 139
Elliot, N., 193
Elliot, R. J., 343
Elliot, R. O., 390
Erbudak, M., 85, 86, 87, 88, 270, 271, 425, 436
Erdös, P., 16, 286, 290, 309, 316, 323, 327, 342, 348, 349, 350, 366, 374, 434

AUTHOR INDEX

Faber, J., Jr., 80, 184, 190, 300
Falaleeva, L. G., 212
Falicov, L. M., 367, 368, 369, 370, 371
Fano, U., 386
Fick, E., 315
Fields, P. R., 245
Fillion, G., 219, 402
Finch, C. B., 227
Fischer, P., 287, 427
Fisk, Z., 45
Flotow, H. F., 82
Forte, M., 285
Fournier, J. M., 95, 126, 152, 263, 279, 285, 286, 296, 388, 416, 434, 435
Fradin, F., 94, 235
Franceschi, E., 393
Franz, W., 153
Frauenfelder, H., 37
Frazer, B. C., 11, 201
Fredo, S., 424
Freeman, A. J., 23, 80, 307, 310, 311, 355, 356, 444
Frick, B., 445
Friedel, J., 364
Friedman, F., 19, 114
Fuggle, J. C., 277
Fulkerson, W., 65, 398
Furrer, A., 288

Gabuda, S. P., 212
Gagarinskii, Yu. V., 212
Gal, J., 154, 280, 424
Galleani d'Agliano, E., 353
Gardner, W. E., 76
Genet, M., 239, 240, 291
Gerward, L., 413
Gibinski, T., 125
Gibney, R. B., 89
Glotzel, D., 359
Godfrey, T. G., 398
Goldman, J., 197
Gonçalves da Silva, C. E. T., 336, 370, 371
Gordon, J., 183
Gortsema, F., 66
Gossard, A. C., 145
Gotoo, K., 180
Greuter, F., 85, 86, 87, 400
Griffin, C. D., 305
Grobler, N. J. S., 448
Grohs, H., 299

Grønvold, F., 116, 122, 179
Gruber, J. B., 233, 242, 243, 325, 326
Gruen, D. M., 169
Grunzweig-Genossar, J., 19, 83, 97, 114, 148, 253, 372, 373
Guillamont, R., 240, 291
Gulick, J. M., 5
Güntherodt, H. J., 400
Gurtovoi, K. G., 421

Hadari, Z., 154
Haessler, M., 127
Haines, H. R., 294
Haire, R. G., 31, 428
Haley, S. B., 374
Hall, R. O. A., 294, 297
Hamaguchi, Y., 161
Hargreaves, W. A., 222
Harvey, A. R., 57, 90, 91, 146, 248
Hauser, E., 400
Heaton, L., 53, 202
Hecht, H. G., 325, 326
Heintz, L., 417
Hellwege, K. H., 321
Hendricks, M. E., 215, 289
Henkie, Z., 32, 106, 107, 134, 136, 137, 138, 422, 437, 438
Hery, Y., 127
Hessler, J. P., 301, 433
Hill, H. H., 178, 298
Hirst, L. L., 380
Höchst, H., 299
Hoekstra, H. R., 403, 403
Hoff, H. A., 276
Holden, T. M., 71, 265
Holland-Moritz, E., 150
Holley, C. E., 89
Holtzberg, F., 20, 397
Horn, S., 150
Horyn, R., 411
Hsu, F., 165
Huang, K., 124, 387
Hubbard, J., 375
Huber, J. G., 17, 157, 171, 172, 177
Hubermann, B. A., 370
Hubert, S., 240, 291
Hufner, S., 299
Hulliger, F., 28, 268, 277, 425, 436
Huntzicker, J. J., 406
Huang, C. V., 384

Huray, P. G., 428
Hussonais, M., 240
Hutchings, M. T., 338
Hutchinson, C. A., Jr., 224

Imoto, S., 115
Irkhin, Y. P., 377
Izyumov, I. A., 35

Jaccarino, V., 145
Jayaraman, A., 394
Jensen, J., 292
Johnston, D., 225
Jones, E. I. D., 339
Jones, E. R., Jr., 213, 215, 289
Jones, W. M., 183
Joos, G., 315
Judd, B. R., 317, 320, 330
Jullien, R., 353, 354, 358
Junkinson, A. R., 8

Kaiser, A. B., 149
Kalvius, G. M., 132, 203, 280, 424
Kamimoto, M., 56
Kanellakopulos, B., 220
Kang, J., 323, 327
Kanter, M. A., 42
Karraker, D. G., 215, 289
Kasper, J. S., 104
Kasuya, T., 422, 423
Kazmierowicz, C. W., 42
Keller, C., 30
Keller, J., 87, 88, 271
Kimball, J. C., 367, 368
Kittel, C., 341
Klamut, J., 137
Klitowski, Z., 6, 107
Knapp, G. S., 278
Knott, H. W., 276
Koehler, W. C., 247
Koelling, D. D.,355, 356, 444
Kollie, T., 398
Komura, S., 161
Konczykowski, M., 437
Konev, V. N.,143
Koster, G. F., 318, 378
Kouvel, J. S., 104
Kruger, O. L., 41, 48, 94
Kruglov, A. A., 143
Krupa, J. C., 240, 291

Krusius, M., 172
Küng, M., 438
Kunii, S., 422
Kunitomi, N., 161
Kutaiber, V. I., 143
Kuznietz, M., 1, 10, 19, 51, 68, 73, 81, 84, 96, 97, 114, 168, 246, 424
Labedev, I. G., 143
Lagnier, R., 221, 262
Lagutin, A. S., 421
Lallement, R., 44, 92
Lam, D. J., 90, 91, 94, 142, 155, 200, 203, 235, 248, 249, 275, 280, 313, 334, 403, 404
Lander, G. H., 2, 4, 10, 33, 40, 51, 52, 57, 60, 69, 70, 71, 78, 80, 90, 91, 103, 150, 159, 184, 246, 249, 260, 265, 267, 273, 275, 276, 297, 300, 335, 412, 449
Lang, J. K., 429
Lasser, R., 277
Lastowski, R. J., 384
Laubschat, C., 281, 282
Lawson, A. C., 177
Lea, K. R., 312
Leask, M. J. M., 185, 210, 312
Leciejewicz, J., 3, 60, 61, 98, 101, 105, 109, 110, 112, 118, 260
Lee, J. A., 294
Lemmer, R. H.,332
Lenander, C. J., 308, 328
Leung, A. F.,217
Levitin, R. Z., 421
Lewis, W. B., 224, 405
Lewityn, R. E., 138
Liberman, D. A., 405
Lin, S., 162
Lines, M. E., 339
Lister, M. W., 186
Loewenhaupt, M., 150, 412
Long, C., 1, 13, 351
Long, E. A., 183
Longworth, G., 81
Lorenzelli, R., 43
Lowther, J. E., 332
Luengo, C. A., 171, 172

McCallum, R. W., 172
McDonald, P. F., 236, 237
McElroy, D., 398
McGuire, T. R., 20

AUTHOR INDEX

Maglić, R. C., 52
Malek, C. K., 291
Manes, L., 279, 296
Mann, J. B., 405
Maple, M. M., 17, 157, 172, 177
Markowski, J. P., 32, 422
Marples, J. A. C., 9, 68, 77, 79
Martin, D. G., 295
Martin, J. F., 8, 46
Martynova, L., 74
Maslanka, R., 437
Meaden, J. T., 389
Meerovici, B., 168
Mehlhorn, R. J., 226
Meier, F., 87, 270
Melamud, M., 166
Methfessel, S., 20
Meyer, H. C., 237
Misiuk, A., 163, 170
Miwa, H., 344
Mollet, H. F., 226
Monachesi, P., 452
Moore, J., 398
Mortimer, M. J., 293, 297
Moser, J., 280
Moser, J. B., 41, 48
Moulton, W. G., 1, 5, 12
Mueller, M. H., 2, 4, 40, 52, 53, 57, 69, 78, 90, 91, 164, 248, 249, 276
Mukaibo, T., 56
Mulak, J., 133, 163, 170, 221, 251, 401
Muller, W., 26, 34, 220, 410
Murani, A. P., 50, 412
Murasik, A., 3, 60, 61, 98, 101, 105, 109, 112, 128, 287, 288, 411, 418
Muromura, T., 64
Murray, A. F., 265
Murray, R. B., 208
Mydlarz, T., 121, 140

Naegele, J. R., 279
Naito, K., 180
Namura, S., 180
Narasimhan, K. S. V. L., 174
Nave, S. E., 428
Nellis, W. J., 18, 57, 156, 147
Nereson, N., 250
Nicol, A. T., 233
Nielson, C. W., 318
Niewenhuys, G. J., 165

Noel, H., 402, 414
Norton, P. R., 264
Nowik, I., 40, 91, 154, 155, 249

Oelhafen, P., 400
Ofer, S., 154
Ogilvie, R., 162
Olcese, G. L., 393
Olsen, C. E., 11, 247, 250, 384, 390, 413
Osborne, D. W., 82, 182, 210, 211
Ozerov, R. P., 35
Ozhogin, V. I., 421

Palewski, T., 3, 61, 98, 101
Pappalardo, P. G., 245
Parks, S. I., 12
Pascard, R., 92
Paszek, A. P., 303
Pauthenet, R., 111
Pechennikov, A. V., 74, 139
Pellegrini, L., 279
Penney, T., 397
Perio, P., 199
Peterson, J. R., 31
Petrov, P. N., 143
Pickett, W. E., 444
Pletyushkin, V. A., 252
Pop, I., 174
Popow, J. F., 138
Potzel, W., 280, 424
Prather, J. L., 396
Pravecki, E., 130
Price, C. E., 135
Pryce, M., 238
Przystawa, J., 108, 113, 129, 130, 131
Ptasiewicz-Bąk, H., 118

Quezel, S., 254, 259, 267, 412, 427, 434, 447

Racah, G., 386
Rahn, L. A., 191
Rajnak, K., 228, 319, 383
Ramirez, R., 268, 369
Rao, V. U. S., 140
Raphael, D., 141
Rauh, E. G., 49
Ray, I., 26
Razafimandimby, H. A., 316, 443
Rebizant, J., 285, 296

Reddy, J. F., 2, 40, 57, 276
Reed, W. A., 165
Reihl, B., 85, 86, 87, 400
Reimer, V. A., 176
Reisfeld, M. J., 232, 234
Reynolds, R. W., 227
Richardson, R. P. 242, 243
Rijkaboer, C., 410
Riley, J. D., 338
Rivier, N., 148, 362
Roberts, L. D., 208, 196
Roberts, L. E., 185
Robinson, J. M., 16, 342, 366, 382, 385, 391, 392
Rodbell, D. S., 15, 340
Rose, M. E., 331
Ross, J., 200
Rossat-Mignod, J., 103, 247, 254, 259, 262, 410, 412, 434, 447
Roudalt, E., 416
Roult, G., 206, 207
Rudermann, M. A., 334
Rudra, P., 309
Rustichelli, F., 296

Sakamoto, M., 161
Salmon, P., 95, 126
Sampson, C. F., 68
Sarup, R., 301
Sandenaw, T. A., 89
Satten, R. A., 223, 225
Scarbrough, J. O., 65
Schenck, A., 39
Schinkel, C. J., 58
Schlabitz, W., 153
Schneider, W. D., 281, 282
Scholten, O., 93
Schönes, J., 192, 266, 272, 415, 430, 438, 445, 446
Schopper, H., 321
Schreiber, C. L., 223
Selwood, P., 195
Shaked, H., 166
Shalimoff, G. V., 301
Shamir, N., 166
Shapiro, S. M., 335, 395
Sharma, K. K., 231
Shavishvili, T. M., 252
Siemann, R., 439
Slichter, C. P., 36
Slinowski, E., 193
Smith, J., 384

Smith, J. L., 178
Smith, T. F., 76
Smotvitskaya, E. S., 153
Solt, G., 286, 290, 348, 349, 350
Sommers, J. A., 116
Soulie, E., 218
Sparlin, D. M., 69
Spirlet, J. C., 26, 33, 34, 279, 280, 285, 410, 413
Squires, G. L., 451
Stalinski, B., 132
Steenstrup, S., 413
Steglich, F., 153, 277
Steiner, P., 299
Steinitz, M., 253
Sterling, W. G., 70, 333
Sternheimer, R. M., 322
Stettler, J. D., 237
Stevens, K. W. H., 314
Stevenson, R., 376
Stone, J., 213
Stone, J. A., 38, 215, 289
Struebing, V. O., 178
Suski, W., 22, 75, 105, 112, 113, 121, 123, 125, 133, 140, 274
Suzuki, T., 422
Svensson, E. C., 71, 265
Szcepaniak, W., 287, 288

Tagawa, H., 64
Takahashi, Y., 56
Takegahara, K., 423
Tapping, R. L., 264
Tchapoutian, R., 262
Tetenbaum, M., 99
Thayamballi, P., 441
Thorn, R. J., 49
Trainor, R. J., 156, 167, 278
Troć, R., 6, 21, 58, 60, 61, 98, 101, 105, 109, 110, 117, 128, 132, 260, 261, 334
Trzebiatowski, W., 21, 75, 106, 138, 195, 216, 256

Unstead, P. B., 47

van Aken, P. B., 173
van Daal, H. J., 144, 173
van Delden, D. C., 448
van Doorn, C. F., 24, 67, 448, 450
van Maaren, M. H., 173
van Vleck, J. H., 379

AUTHOR INDEX

Varga, L. P., 232, 234
Veal, B. W., 284, 403, 404
Vettier, C., 262, 263
Vogelsang, W., 54
Vogt, O., 28, 62, 69, 70, 102, 103, 254, 257, 258, 259, 262, 265, 266, 267, 268, 273, 335, 400, 415, 418, 445, 447
Voigt, W., 387
von der Lage, F. C., 324
Vonsovskii, S. V., 381

Wachter, P., 257, 409
Walker, C. T., 187, 191
Wallace, W. E., 174
Wagner, E. F. E., 255
Walter, A. J., 185
Wang, Y. L., 13
Warren, I. H., 135
Warren, J. B., 39
Watson, R. E., 307, 310, 311
Watters, R. R., 82
Weber, L. W., 164
Wedgwood, F. A., 50, 68, 72, 73, 81, 263, 343
Weger, M., 166
Weling, F., 452

Wernick, J. H., 145
Westrum, E. F., Jr., 70, 79, 82, 116, 182, 211, 406
Williams, J., 202
Williams, G. S., 52
Wohlfarth, E. P., 256
Wohlleben, D., 17, 157
Wojakowski, A., 125, 416, 417, 420, 434
Wolfers, P., 219, 402
Wolf, W. P., 185, 210, 312
Wong, E. Y., 223, 225
Wortmann, G., 280
Wybourne, B. G., 230, 319

Yanase, A., 423
Yang, D., 440, 442
Yarembash, E. I., 74, 139
Yessik, M., 27

Zaki, M. R., 116
Zauberis, D. D., 53
Zlatić, V., 362
Żołnierek, Z., 117, 221, 251, 286, 350, 401, 426
Zuckermann, M. J., 148
Zygmunt, A., 108, 118, 119, 120

Subject Index

(Main entries in *italics*)

A

Actinide laboratories, 7
   metals, *149–152*
Anderson model, 4
Anderson superexchange, *123*, 127
Anisotropic exchange, 123, *124*, 155
Anisotropy, magnetic, 23, 31, 33, 36, 48, 50, 54, 58, 67, 91, 132, 142
Antiferromagnetic resonance, 78
Antiphase structure, 45

Am, 2, 84, 149, *150*, 151
Am^{2+}, 91
Am^{3+}, 68, 91, 95, *96*, 97
$AmCl_3$, 96
AmF_3, 91
$AmFe_2$, 67, *68*, 169
$Am(C_5H_5)_3$, 91
AmO_2, *84*
Am_3Se_4, *61*
$AmTe_4$, *61*
$AuCu_3$-structure, 62, 71

B

Band antiferromagnetism: see Itinerant antiferromagnetism
Band models, 4, 67, 101, *149–155*
Band structure calculations, *149–155*
Biquadratic exchange, *125*, 133
Bremsstrahlung isochromat spectroscopy, 13

Brillouin curve, 1, 34, 37, 43, 44, 61, 67, 90, *133*
   zone, 140, 153
   zone boundaries, 28, 32

Bk, 2, 149, 150, 152
BkO_2, *84*

C

Casimir operator, 107
Chemical vapor transport, 8
Classes, of actinide compounds, 1
Clebsch–Gordan coefficients, *116*, 172
Clustering, 80
Conduction electron spin polarizations: see Spin polarization
Conductivity: see Electrical resistivity
Conductors, 52
Configuration interaction, 105
Coqblin–Schrieffer exchange, *129–133*
Correlation, electronic, 4, 103, 104, *164*, 165
Coulomb interactions: see Slater integrals (or parameters)
Covalency, 15, 90, *104*
Critical exponents, 36
Critical neutron scattering, *3*, 22, 76
Crystal-field model, 4, 43, 50, 57, 59, 71, 89, 92, *99–123*, 142
   field parameters, 91, 92, 93, 95, 97, *112*, 133
   field splittings, 104

Crystal-field model (*cont.*)
  field theory, 81, *92*
  symmetry, 113–115
Curie–Weiss constant, 21, 77, 80
  law, 15, 27, 28, 41, 55, 62, 73, 86, 88, 90, 153

C15-type Laves structure, 67
$CaCl_3$, 109, 122, 124, 127
$CaCu_5$, 73
$CaAl_2Si_2$, structure, 74
CaF_2, lattice or host, 68, 74, 92, 93, 94, 111, 140
Ce, 150, 168
Ce^{3+}, 124, 129, 130, 132, 168
$CePd_3$, 169
Cf, 2, 149, 152
Cm, 2, 149, 150, 152
Cm^{3+}, 95
CmF_4, 91
CmO_2, 84
$CoUS_3$, 91
CrB-structure, 72
$CrUS_3$, 91
$CrUSe_3$, 91
$CsUF_6$, 91
Cs_2ZrCl_6, host, 93, 96
Cu_2Sb, structure, 51

D

Debye temperature, 27
Defect structure, 42
Delocalization, of electrons, 66, 72, 90, 150, 156, *157–163*
Density of states, electronic, 13, 26–27, 35, 72, 153, 154
Dielectric function, 31
Dipolar interaction, 90
Dirac equation, 104, 154
Dispersion curves, see Phonons; Spin waves
Distortion: see Lattice distortion
Domains, antiferromagnetic, 8, 140
Double-k structure: see Two-k structure
Doublet, 72

E

Easy axis, 18, 58
Effective field, 133, 146
Elastic constants, 29, 30, 77, 139
Electric multipole interactions, 125

Electrical resistivity, 13, 17, 19, 22, 28, 32, 33, 36, 37, 39, 40, 41, 42, 43, 44, 47, 55, 56, 61, 62, 66, 67, 68, 69, 70, 71, 145, 157, 161, 167
  magnetic part, 59
Electron–phonon interaction, 162
Entropy, 15, 17, 26, 27, 53, 59, 72, 76, 83, 86, 88, 160
Exchange
  anisotropic, 123, *124–125*, 132, 155
  biquadratic, 125, *133–135*
  Coqblin–Schrieffer, 129–133
  Heisenberg, 4, 36, *122–126*, 133
  indirect, 90
  interaction, 53, 89, 91
  RKKY, *125*, 129; see also RKKY interaction
  striction, 135

Es, 2

F

Fermi liquid, 66
Ferrimagnetism, 23, 91
Ferromagnetism, 4, 23, 37, 67, 127
5f electrons: see Localized moments
First-order transitions, *1*, 4, 20–23, 41, 53, 70, 71, 74, 76, 87, 133–148, 157–163
Fluorescence IPE spectroscopy, 13
Form factor, 71, 144, 180
Fourier analysis, 176
Fractional parentage, 165

Fe, 2
Fm, 2

G

Galvanomagnetic properties, 34, 154
Green's function, 166

Gd, 2

H

Hall coefficient, 59, 61
Hamiltonian
  of actinide atoms, 164
  of band electrons, 152
  Coulomb, 102, 105, *106–108*
  crystal-field, 102, 105, 148, 172

SUBJECT INDEX

Hamiltonian (*cont.*)
 of 5f electrons, *102*, 135, 136, 145
 magnetic, 102
 spin orbit, 102, 105, *108–111*
 spin–other-orbit, 108
 spin–spin, 108
Heat capacity: see Specific heat
Heisenberg exchange: see Exchange
High-pressure experiments: see Pressure experiments
Hill's rule, 72
Hubbard model, 4, *166*, 167
Hund's rule, 72, 91
Hybridization, 31, 43, 74, 129, *149–154*, 165–167
Hyperfine field, 12, 34, 66, 83, 96, 163
Hysteresis, magnetic, 18, 23

I

Incommensurate structure, 38–39, 50, 175, 176
Induced moment systems, 56, 90
Inelastic neutron scattering, 15, 23, 29, 31, 36, 66, 71, 78, 88
Insulator, ferromagnetic, 85
Intermediate valence, 29, 31, 34, 50, 63, 68, 73, 101, 123, 163, *169–170*
Intermetallics, 4, 63, 65, 66, 155, 167
Iron-group metals, 4
Irreducible tensor operators, *171–173*
Ising model, 36
Isomer shift, *10*, 61, 66, 167
Itinerant antiferromagnetism, 26, 72, 153, 154, 155
Itinerant ferromagnetism, 66, 67, 73
Itinerant states, 4, 44, 63, 66, 68, 72, 100, 130, *148–168*

I_2, 8

J

Jahn–Teller effect, 92

K

Knight shift, 73
Kramers doublet, 4, 96
Kramers–Kronig relations, 31
Kubic harmonics, 113

L

Lattice constants, 19, 20, 26, 40, 42, 80
 distortion, 29, 33, 35, 37, 39, 40, 43, 51, 55, 61, 78, 79, 82, 86, 90, 135, 136, 140–142
 vacancies, 82
Lea, Leask, and Wolf parameters, *96*, 122, 145
 diagram, 104
Localized moments, 35, 42, 44, 152, 153, 156
Localized states, 4, 11, 39, 63, 66, 67, 72, 84, 100, 130, 150, 158–163, *164–169*
Longitudinal magnetization wave, 38, 47

$LaBr_3$, 95, 96, 124
$LaCl_3$, host, 95, 96, 97, 98, 109, 122, 124
$LiNO_3$, host, 96
Lr, 2

M

Magnetic excitations, 23, 29, 36, 139, 140; see also Spin waves
Magnetic moment, 21, 26
Magnetic ordered moment
 rotation of, 17
 see also Ordered moment
Magnetic ordering, types of, *15*, 38, 53, 127–133
Magnetic phase diagram, 21, 24, 40, 45, 46, 48, 49
Magnetic susceptibility, 3, 12, 15, 17, 18, 22, 23, 27, 28, 34, 35, 37, 40, 42, 46, 54, 59, 62, 63, 66, 68, 69, 71, 80, 81, 82, 83, 86, 90, 91, 92, 139, 142, 148, 161
Magnetic symmetry, 3, 126, 145, 162
Magnetization measurements, 12, 33, 34, 43, 46–50, 68, 134
Magnetization wave, longitudinal, 38, 47
 sinusoidal, 57
Magnetoelastic coupling, 89, 125
Magnetostriction, 135
Magnon–phonon coupling, 77, 78
Marvin integrals, 107
Melting points, 15
Metal–insulator transition, 158
Mixed valence: see Intermediate valence
Mixing
 p-f, 59
 s-f, 166
Molecular field, 133, 135, 146

SUBJECT INDEX

Moment-jump transitions, *1*, 3, 4, 19–20, 38, 40, 46–47, 85, 144, 157–163
Mössbauer resonance, *9*, 34, 57, 61, 66, 67, 68, 72, 83, 84, 85, 142
Mott transition, 35
Muffin-tin potential, *150*, 152
Multiaxial structure, 50
Multiplet, *108*, 109, 110
  mixing, 108, 122
Multipole, 125
Muon spin rotation, 11–12
Mystery, 82

Md, 2
MgCu$_2$-structure, 62

N

Neutron diffraction, *8*, 16, 22, 23, 26, 31, 33, 37, 41, 44, 50, 53, 57, 70, 71, 76, 78, 83, 87, 89, 90, 91, 122, 132, 142, 146, 163, 177, *179–180*
  see also Inelastic neutron scattering; Polarized neutron diffraction
Nonmagnetic ground state, 68
Nuclear magnetic resonance, *9*, 16, 57, 87, 90, 145, 148, 163

NaCl-structure, 13, *15*, 127, 132, 153, 154
Nd$_{1-x}$U$_x$Co$_5$, 73
NiUS$_3$, *91*
Np, 2
Np ions, 9, 66, 72, 78, 143
Np metal, 1, 2, 149, *150*, *156*
Np^{3+}, 95, *96*, 97, 98, 122, *142*, 143, *144*
Np^{4+}, 84, *96*, 142, 143, 144
NpAl$_2$, 3, 9, 64, *66*
NpAs, 14, 39, *40*, 41
NpAs$_2$, *57*
Np$_3$As$_4$, *61*
NpC, 1, 3, 14, *36*, 37, 38, 92, 127, 161
NpCl$_4$, 86, *90*
NpCo$_2$, 64, 67
NpF$_6$, 93
NpFe$_2$, 64, *67*
NpH$_2$, *68*
NpIr$_2$, 64, *66*
NpN, 3, 14, *41*
NpNi, 64, *66*
NpO$_2$, 3, 7, 74, 75, 77, 78, *83*, 84, 139, *142*, 143, 144

NpOs$_2$, 9, *66*, *167*, 170
NpP, 13, 14, 15, *38*, 39, 47
NpPd$_3$, 3, *71*, 72
NpS, 14
NpSb, 14, *40*, 42
NpSn, 65, *71*, 72

O

One-dimensional ordering, 88
Optical reflectivity, 31, 79
Optical spectroscopy, 35, 91, *92–98*, 101, 154
Orbital moment, 129, 132
Order–disorder transition, 143
Ordered moment, 15, 40, 42, 43, 44, 46, 53, 66, 83, 87, 137

P

Paramagnetic resonance, 91, 96
  susceptibility: see Magnetic susceptibility
Paramagnons, 66, *155*, 157
Phase diagrams: see Magnetic phase diagrams
  transitions: see First-order transitions; Second-order transition
Phonons, 15, 23, 29, 36
Photoelectron spectroscopy: see Photoemission
Photoemission, 5, 12, *13*, 29, 34–35, 44, 66, 67, 73, 84, 85, 169
Point-charge calculations, *121*, 148
Point symmetry, *113–115*, 148
Polarized neutron diffraction, 67, 84
Pressure experiments, 4, 23, 25, 26, 27, 35, 66, 67, 72, 169

Pa, 2
PaAs$_2$, *57*
PaCl$_4$, *90*
PbFCl-structure, 51
Pm^{3+}, 95
Pu, 1, 2, 149, *150*, *156*, 157
Pu^{3+}, 95, 97, 98
PuAl$_2$, *62*, 64
PuAs, 14, *44*
PuC, 3, 14, *42*, 43, 44, 154
PuFe$_2$, *67*
PuH$_2$, *68*
PuH$_3$, *72*
PuIr$_2$, 64
PuN, 14, 42

SUBJECT INDEX

PuO_2, *84*
PuP, 14, *43*, 44, 154
$PuPd_3$, 65
$PuPt_2$, *62*, 64
$PuPt_3$, 65, 69, 72
$PuRh_2$, 64, *68*, 69
$PuRh_3$, 65
$PuRu_2$, 64
PuS, 13, 14, *44*, 154
Pu_3S_4, *61*
PuSb, 14, *44*

Q

Quadrupolar interactions (or moments), 123, 125, *135*–145, 148
Quadrupole resonance, 5

R

Racah operators, *116*, 171–173
   parameters, 91, 96, *106*, 110, 116
Radial integrals, 116, 121
Radioactivity, 7
Raman scattering, 79
RAPW method, 150
Rare earths, 4, 99, 124, 169
   intermetallics, 63
Reciprocal lattice, 177–183
Relativistic calculations, 104, 150, 154
Resistivity: see Electrical resistivity
RKKY interaction, 48, *125*–*129*, 161, 162
Russell–Saunders states, 90, *110*

S

Sample preparation, 7–8
Saturation moment, 15, 61, 72
Schottky anomaly, 90
Screening effects, 17, 44
Second-order transition, 3, 89, 138
Semiconducting energy gap, 55
Semiconductors, 13, 44, 52, 55, 84, 91
Semimetals, 33, 59
Seniority, 106
Shielding, 112, *120*–*121*, 148
Short-range order, 77, 89
Single crystals, 7–8
Singlet ground state, 56, 78, 81, 146, 162
Slater integrals (or parameters), 92–98, *106*, 111, 117
   determinants, 106

Solid solutions, 15, 21, *44*–*51*, 74, 79, 128
Sound velocity, 30
Specific heat, 3, 13, 23, 25, 26, 27, 34, 37, 42, 44, 47, 48, 53, 54, 66, 69, 71, 72, 76, 82, 83, 86, 87, 90, 142, 153, 162
Spectroscopic parameters, 91, 96
Spherical harmonics, *111*, 113, 115, 121, 131, 148
Spin-disorder scattering, 28, 39, 41, 71, 155
Spin fluctuations, 62, 63, 66, 71, *155*–157, 167
Spin–lattice interaction, 93, 135
Spin–orbit interaction (or coupling), 91–98, (75), 105, 106, *108*–111, 150
Spin polarization, 12, 33–34, 35, 43
   resonance, 12
   rotation (or reorientation, or canting), 16, 73, 85, 86
   waves, 23, 31, 36, 59, 67, 73, 78, 155
Standard basis operators, 132, *164*–169
Sternheimer shielding factors, *112*, 120
Stevens operators, 121
Stoner enhancement, 157
   excitation, 67
Sublattice magnetization, 16, 22, 23, 26, 28, 29, 30, 71, 76, 146, 148, 161

SmS, 169
$SmY_{1-x}S_x$, 169
Sn, 72

T

Tensor operators; see Irreducible tensor operators
Ternary compounds, 55, 57, 91
Tetragonal structure, 52, *53*, 56, 91, 129
Thermal conductivity, 77
Thermal expansion, 42, 161
Thermal resistivity, 66
Thermodynamic functions, 117
Thermoelectric power (or thermopower), 13, 42, 43, 44, 60, 66
Tight-binding, 151, *165*
Three-k structure, 31, 33, 50, *51*, 133, 175, 183
Two-dimensional ferromagnet, 74
Two-k structure, *17*, 22, 132, 133, 175, 183

$TeCl_4$, 8
Th, 2, 74, 75, *149*, 150
$ThBr_4$, 92
ThC, 43

SUBJECT INDEX

$ThCl_4$, 93
ThO_2, 76, 77, 79, 80, 81, 84, 85
ThP, 21, 80
Th_3P_4, structure, 58
$ThPd_3$, 69, 71
Ti, 74
TmSe, 123, 169

U

Ultrasonic paramagnetic resonance, 93

U ions, 51, 133
U metal, 1, 2, 74, 149, 150, *151*
U^{3+}, 50, 88, 89, 93, *95*, 97, 98, 107, 109, 112, 120, 148, 159, 160, 169
U^{4+}, 50, 56, 57, 71, 78, 90, 92, 93, *94*, 96, 112, 116, 118, 120, 124, 135, 139, 145, 159, 160
UAl_2, *62*, 64, 66, 156
UAl_3, 65
UAl_4, *166*, 167
UAs, 1, 3, 7, 8, 14, 16, 20, *22-26*, 46, 101, 122, 128, 133, 145, 152, 154, 161, 162, 163, 170
UAs_2, 52, *53*, 54, 129
U_3As_4, 58-61
$UAs_{1-x}P_x$, 44, *45*, 161
UAsS, 52, *54*
$UAs_{1-x}S_x$, 15, 44, *47*, 48, 128, 162
UAsSe, 52, *54*
$UAs_{1-x}Se_x$, *47*, 49, 128
UAsTe, 51, 52, *54*
UBi, 14
UBi_2, *52*, 129
U_3Bi_4, 59
UBr_3, *86*, 88
UBr_4, 86
UC, 14, 43, 44, 152
UCl_3, 86, 88, *89*
UCl_4, 86, *90*, 91, 93, 117
UCl_6^{2-}, *92*
UCu_5, *73*, 168, 169
UCu_2As_2, *73*
UCu_2P_2, *73*
$UCrO_4$, *85*
UD_3, 65, *72*
UF_4, 86, *90*
UFe_2, 7, 64, *67*
$UFeO_4$, 75, *85*, 175
UGa_2, 64, *68*
UGa_3, *65*, 155

UGe_2, 64
UGe_3, 65
UGeY (Y = S, Se, Te), *54*
UH_3, 7, 8, *65*
UHg_2, 64
UI_3, 3, 7, *86-89*, 146, *147-148*
UIn_3, 65, *72*
UMn_2, 64, *68*
UN, 7, 14, *26-30*, 51, 122, 123, 127, 152-154, 170
U_2N_3, *91*
UNi_5, 73
$UNi_{5-x}Cu_x$, 73, 168, 169
UNSe, 52, 57
UNTe, 52, 57
U_2N_2X (X = P, As, S, Se, Sb, Bi, Te), 3, *91*
UO_2, 3, 7, 8, 21, 71, *74-84*, 87, 93, 124, 125, *135-142*, 143, 145, 146, 147
U_3O_7, 75, *82*, 91
U_3O_8, 75, *82*, 91
U_4O_9, 75, *82*, 91
UOS, 3, 52, *54*
UOSe, 52, *55*, 56
$UOSe_2$, 56
UOTe, 52
UP, 1, 3, 4, 7, 8, 9, 14, *16-22*, 23, 26, 44, 45, 46, 52, 80, 123, 125, 127, 129, *133*, 144-146, 152, 153, 158-163, 169
UP_2, 3, 7, 8, *55*, 56, 129
U_3P_4, 3, 7, 8, 53, *58-61*
UPd_3, *69-71*
UPS, *46*, 162, 169
$UP_{1-x}S_x$, 15, *45-47*, 48, 128, 154, 161
$UP_{1-x}Se_x$, 44, *47*, 49, 128
UPt, *72*, 73, 167, 169, 170
UPt_2, *73*
US, 14, *33*, 34, 35, 36, 37, 44, 45, 127, 152, 154, 162, 163, 170
US_2, 51, 52, 57, 58
U_2S_3, *91*
USb, 7, 14, *31*, 33, 50, 51, 122, 127
USb_2, 52, *53*, 54, 129
U_3Sb_4, 59, *61*
USbTe, *50*, 51
$USb_{1-x}Te_x$, *50*
USe, 14, 21, *33-36*, 44, 127, 152
USe_2, 51, 52, 56
U_3Se_4, 59, *61*
USi_3, 65
USn_3, 65, 156

SUBJECT INDEX

UTe, 14, *33*–36, 127
UTe$_2$, 56
U$_3$Te$_4$, 59, *61*
UThCl$_4$, *90*
U$_x$Th$_{1-x}$Cl$_4$, 90
U$_x$Th$_{1-x}$O$_2$, 21, 78, *79*–81
U$_x$Th$_{1-x}$P, 80
U$_x$Th$_{1-x}$S, 48, 154
U$_{1-x}$ThS$_x$, *47*
UThSb, 51
U$_x$Th$_{1-x}$Sb, *48*, 50
UTl$_3$, 65, *72*

V

Valence transitions: see Intermediate valence
Van Vleck formula, 144
Vaporization experiments, 15
Vibronic spectra, 12, *93*

Virtually bound states, 47, 63, 154, 155, 164
Volume changes, 3, 20, 39–40, 77, 135, 163

VUS, 91

W

Wigner-Eckart theorem, *171*, 173
Wigner n–j symbols, *116*, 171, 173

X

X-ray diffraction, 8, 12, 82
  photoelectron spectroscopy: see Photoemission

Z

Zeeman hamiltonian, 57, 117, 120, 147
  spectroscopy, 97